Telescope Eyepieces

Anyone who has looked through binoculars, a telescope or microscope has used an eyepiece. *Telescope Eyepieces: Optical Theory and Design* explores the wide range of eyepiece designs. It introduces optics theory progressively to build understanding of how lenses control light in an optical system, both generally and in eyepieces specifically, linking optics fundamentals, design evolution and the implications for image quality. This book presents a logical narrative starting with Snell's law of refraction at a flat surface, progressing to paraxial and real rays at spherical and aspherical surfaces, lenses and thin-lens systems. It demystifies the origin of aberrations by considering wavefront deviations, all brought to life in the context of the familiar eyepiece. Principles are explored both descriptively and mathematically, and carefully interpreted so the reader is not swamped by a sea of equations. This book contains many diagrams of ray tracing results to illustrate optical principles and the consequences of design choices, enabling the reader to visualise their impact on image quality and to cut through the puffery that is sometimes found in the marketing of telescope eyepieces. It bridges the gap between introductory books and expert-level optical-design texts, written jointly for an astronomy readership and for physics and optical-design students.

Key Features:

- Develops an understanding of optics by focusing on the design variations of a particular multi-lens system – the eyepiece – which readers will have used when looking through binoculars, telescopes and microscopes.

- Combines descriptions and interpretations of theory with many illustrative visualisations of eyepiece designs to help the reader develop an intuitive understanding of optics and aberrations in the familiar context of eyepieces.

- Written by a Professor of Astrophysics with experience as both an amateur and professional observer using telescopes at observatories around the world.

Sean G. Ryan is a professional astronomer with almost 50 years of experience as an amateur observer. He was appointed Professor of Astrophysics at the University of Hertfordshire in 2006, where he was Head and Dean of the School of Physics, Astronomy and Mathematics for ten years. He has published over 100 research papers on observational astronomy and has co-authored several books, including *Visual Astronomy with a Small Telescope.*

Telescope Eyepieces
Optical Theory and Design

Sean G. Ryan

CRC Press
Taylor & Francis Group
Boca Raton   London   New York

CRC Press is an imprint of the
Taylor & Francis Group, an **informa** business

Designed cover image: Sean G. Ryan, generated from a ray trace using WinLens Basic

First edition published 2026
by CRC Press
2385 NW Executive Center Drive, Suite 320, Boca Raton FL 33431

and by CRC Press
4 Park Square, Milton Park, Abingdon, Oxon, OX14 4RN

CRC Press is an imprint of Taylor & Francis Group, LLC

ISBN: 978-1-041-13587-6 (hbk)
ISBN: 978-1-041-13254-7 (pbk)
ISBN: 978-1-003-67050-6 (ebk)

DOI: 10.1201/9781003670506

Typeset in Minion
by codeMantra

To the glass workers and optical designers of the past, present and future, for revealing the beauty of the visible universe.

Contents

Acknowledgements

I HAVE LONG BEEN FASCINATED with the ability of a fused mass of sand, sodium carbonate and limestone, i.e. glass, to manipulate the passage of light, and by the exploitation of this capability to form well-focussed images in astronomical observation, photography, spectroscopy and microscopy. As an amateur astronomer in my teens, fellow members of the Canterbury Astronomical Society, especially David Buckley, set me on an optical journey that continued for decades, and which has benefitted from enjoyable discussions with many colleagues, especially my friend Richard Greenaway from whom I learned much. This book could not have been written and presented as envisaged without the ray tracing software WinLens Basic, written by Dr Geoff Adams who kindly agreed to its use for this purpose.

Table of Equations

$$n \sin i = n' \sin i' \tag{2.1}$$

$$V_{\mathrm{D}} \equiv \frac{n_{\mathrm{D}} - 1}{n_{\mathrm{F}} - n_{\mathrm{C}}} \tag{2.2}$$

$$\sin i = \frac{r - l}{r} \sin \theta \tag{2.3a}$$

$$\sin i' = \frac{n}{n'} \sin i \tag{2.3b}$$

$$\gamma = i - \theta \tag{2.3c}$$

$$y = r \sin \gamma \tag{2.3d}$$

$$\zeta = r - r \cos \gamma \tag{2.3e}$$

$$\theta' = i' - i + \theta = i' - \gamma \tag{2.3f}$$

$$l' = r \left(1 - \frac{\sin i'}{\sin \theta'} \right) \tag{2.3g}$$

$$\sin(x) = x - x^3/3! + x^5/5! - x^7/7! + \dots \tag{2.4}$$

$$\frac{n'}{l'} =_{\mathrm{P}} \frac{n}{l} + \frac{n' - n}{r} \tag{2.5}$$

$$F_{\mathrm{surf}} \equiv \frac{n' - n}{r} \tag{2.6}$$

$$\frac{n_2'}{l_2'} =_p \frac{n_1}{l_1} + \frac{n_1' - n_1}{r_1} + \frac{n_2' - n_2}{r_2} \tag{2.7}$$

$$\frac{1}{l_2'} =_p \frac{1}{l_1} + \frac{n_g - 1}{r_1} + \frac{1 - n_g}{r_2} \tag{2.8}$$

$$F_{\text{thinlens}} = F_{\text{surf}\,1} + F_{\text{surf}\,2} \tag{2.9}$$

$$f' = n'/F_{\text{surf}} \tag{2.10}$$

$$f' = 1/F_{\text{thinlens}} \tag{2.11}$$

$$f = -n/F_{\text{surf}} \tag{2.12}$$

$$f = -1/F_{\text{thinlens}} \tag{2.13}$$

$$F_E = F_1 + F_2 - (d/n) F_1 F_2 \quad \text{and} \quad f_E' = 1/F_E = -f_E \tag{2.14}$$

$$F_V' = \frac{F_E}{1 - \dfrac{d}{n} F_1} \quad \text{and} \quad f_V' = 1/F_V' \tag{2.15}$$

$$F_V = \frac{F_E}{1 - \dfrac{d}{n} F_2} \quad \text{and} \quad f_V = -1/F_V \tag{2.16}$$

$$F_E = F_1 + F_2 - d F_1 F_2 \tag{2.17}$$

$$m_{\text{ang}} \equiv \theta'/\theta =_p f_{\text{tel}}'/f_{\text{eye}}' \tag{3.1}$$

$$\zeta = r - \sqrt{r^2 - y^2} \tag{3.2}$$

$$\Delta \zeta =_s A\rho^4 = A_0 \left(y_A^2 + x_A^2 \right)^2 + A_1 h' \left[4 y_A \left(y_A^2 + x_A^2 \right) \right] + A_2 h'^2 \left(6 y_A^2 + 2 x_A^2 \right) \\ + A_3 h'^3 \left(4 y_A \right) + A_4 h'^4 \tag{3.3}$$

$$TA_y = l' \times \left(\delta\zeta / \delta y \right) \tag{3.4}$$

$$LA_y = TA_y \times (l'/y) \tag{3.5}$$

$$\frac{n'}{l'} = \frac{n}{l} + F_{surf} + y^2 \left(\frac{n'}{2l'} \left(\frac{1}{r} - \frac{1}{l'} \right)^2 - \frac{n}{2l} \left(\frac{1}{r} - \frac{1}{l} \right)^2 \right) \tag{3.6}$$

$$\frac{3(2n+1)}{4n} p + \frac{3(n+1)}{4n(n-1)} q \tag{3.7}$$

$$\frac{-1}{r_P} = \frac{1}{n_1 f_1'} + \frac{1}{n_2 f_2'} + \ldots = \frac{F_1}{n_1} + \frac{F_2}{n_2} + \ldots \tag{3.8}$$

$$\zeta = \frac{y^2}{r \left(1 + \sqrt{1 - \dfrac{y^2}{r^2}} \right)} \tag{6.1}$$

$$\zeta = \frac{1}{2} \frac{y^2}{r} + \frac{1}{8} \frac{y^4}{r^3} + \frac{1}{16} \frac{y^6}{r^5} + \ldots \tag{6.2}$$

$$\zeta = \frac{\rho^2}{r \left(1 + \sqrt{1 - \varepsilon \dfrac{\rho^2}{r^2}} \right)} \tag{6.3}$$

$$\zeta = \frac{\rho^2}{r \left(1 + \sqrt{1 - (1+k) \dfrac{\rho^2}{r^2}} \right)} \tag{6.4}$$

$$\zeta = \frac{1}{2} \frac{y^2}{r} + \frac{1}{8}(1+k)\frac{y^4}{r^3} + \frac{1}{16}(1+k)^2 \frac{y^6}{r^5} + \ldots \tag{6.5}$$

$$\zeta = \frac{\rho^2}{r \left(1 + \sqrt{1 - (1+k) \dfrac{\rho^2}{r^2}} \right)} + a_4 \rho^4 + a_6 \rho^6 + a_8 \rho^8 + a_{10} \rho^{10} + \ldots \tag{6.6}$$

Glossary of Variables and Points

ROMAN FONT

a, b	semi-major and semi-minor axes of ellipse and hyperbola
a_4, a_6, a_8, a_{10}	unspecified polynomial coefficients for non-conic aspheric function
a, b, c, A, B, C	arbitrary sides and opposing angles of triangle
a, h, o	adjacent, hypotenuse and opposite sides of right-angled triangle
$A, A_0, A_1 \ldots A_4$	unspecified aberration coefficients in the Seidel analysis
A, A_1, A_2	aperture stop, or arbitrary point in optical path
B	Barlow lens factor
c	speed of light in vacuum
$C_\mathrm{I}, C_\mathrm{II}$	longitudinal and lateral chromatic aberration coefficients (see also S_I–S_V)
C	centre of curvature of a surface
d	distance between two surfaces of a lens or two lenses in a system
D	diameter of telescope entrance aperture
e	eccentricity of a conic section, cf. ε and k
f, f'	first and second focal lengths of a surface or lens
f'_E	effective focal length of a lens or system
$f_\mathrm{V}, f'_\mathrm{V}$	first and second vertex focal lengths of a lens or system
$F, F_\mathrm{E}, F_\mathrm{V}$	focal power, effective focal power or vertex powers of lens or system
F, F'	first and second focal points of a surface, lens or system
g	gap fraction for separation of field lens and eye lens in Ramsden eyepiece

h'	distance of an image point from the optical axis
i, i'	angles of incidence and refraction
k	conic constant; cf. e and ε
l, l'	distance of object and image points from vertex of surface or thin lens, or from principal planes of lens or system, in Cartesian coordinate system
L, L'	object vergence and image vergence
LA	longitudinal aberration
m	angular order of Zernike polynomial; see also Z, R
$m_{\text{lat}}, m_{\text{ang}}$	lateral and angular magnifications
n, n'	refractive index of optical medium in object space and image space
$n_{\text{a}}, n_{\text{g}}$	refractive index of air and glass
$n_{\text{C}}, n_{\text{d}}, n_{\text{D}}, n_{\text{F}}$	refractive indices at wavelengths specified by the subscript labels
n	radial order of Zernike polynomial; see also Z, R
N, N′	first and second nodal points of a surface, lens or system
NA	numerical aperture
O, O′	object and image points of a surface, lens or system
p, q	Coddington position and shape factors for object and lens respectively
P, P′	first and second principal planes of a surface, lens or system
$r, r_{\text{s}}, r_{\text{t}}, r_{\text{P}}$	radius of curvature of a surface, especially sagittal, tangential and Petzval focal surfaces
$R, R_{0°}$	Ratio of eye-lens power to field-lens power in Huygens eyepiece. Fresnel reflectance at 0° incidence
R, R_n^m	radial term of Zernike polynomial of angular order m and radial order n; see also Z and Φ
s, s'	distance of object and image points from vertex of surface or thin lens, or from principal planes of lens or system, in all-positive coordinate system
$S_{\text{I}}, S_{\text{II}}, S_{\text{III}}, S_{\text{IV}}, S_{\text{V}}$	Seidel aberration coefficients for spherical aberration, coma, astigmatism, Petzval curvature and distortion
t	elasped time, or thickness of lens (context makes usage clear)
T	transmission fraction

TA	transverse aberration
v, v'	speed of incident and refracted wavefront
V_D	dispersive power, Abbe number or constringence, at the wavelength of the D spectral line
V, V_1, V_2	vertex of optical surface
x	arbitrary variable, as in sin x. See also x,y,z.
x, y, z	distances measured along any of the three directional axes corresponding to: x) distances behind and in front of the meridional plane, y) above and below the optical axis, and z) along the optical axis. The ray centrations x and y are commonly measured at an optical surface or in the aperture stop, cf. h' which is the measured in the image plane
x_A, y_A, y_P	x and y distances measured explicitly in the aperture stop, y_P being for the principal ray. See x,y,z.
Z, Z_n^m	Zernike polynomial of angular order m and radial order n; see also R and Φ

GREEK FONT

α, α'	(alpha) angle between ray and optical axis in all-positive coordinate system, also arbitrary angle
γ	(gamma) angle between optical axis and normal to a point on a spherical surface
ε	(epsilon) variable describing a conic section, cf. e and k
ζ	(zeta) sag of a surface (wavefront, lens or focal surface) measured along the z-axis from the vertex
θ, θ'	(theta) angle between ray and optical axis in Cartesian coordinate system, also arbitrary angle
λ	(lambda) wavelength
ρ	(rho) radial centration of a ray, cf. x,y
υ	(upsilon) upstream distance of Barlow lens from focal plane of telescope
φ	(phi) angular coordinate of a point in an aperture, in polar coordinates
Φ	(Phi) angular term of Zernike polynomial; see also Z and R

Table of Figures

CHAPTER 1

Introduction

1.1 ROLE OF THE EYEPIECE AND KEY TERMINOLOGY

The purpose of the main optics of a telescope, upstream of the eyepiece, is to gather the light of a distant object and cause the rays of light from each object point to converge at a common point of intersection. An image formed by such *converging* rays is termed a real image, and placing a white card in the position where the image is formed will make it visible, given sufficient light. The same cannot be said for the image of a nearby object, such as an insect, seen through a magnifying glass. In normal use, a magnifying glass directs *diverging* rays of light towards the eye, and the image that the eye sees is termed a virtual image; it cannot be seen on a white card introduced into the beam, in contrast to the case for a real image. The virtual image exists only by virtue of being the apparent point of origin of the diverging rays.

For telescopes used visually, the image formed by the main optics must be projected into the eye of an observer, and at the same time, magnified. This is achieved with a specialised set of optics close to the eye – the eyepiece – which is the principal subject of this book.

It would be possible to design the main optics of the telescope and the eyepiece together as an optimised pair, but it is often preferable for a telescope to be equipped with a number of interchangeable eyepieces that can provide different magnifications and fields of view, and which therefore can be used for different purposes. Consequently, the practical convention is for eyepieces to be designed largely independently of the main telescope. It is for this reason that this book will, as much as possible, ignore the optical details of the telescope and concentrate instead on

DOI: 10.1201/9781003670506-1

the designs of various eyepieces that may be employed, each possessing different advantages and disadvantages.

A telescope is typically specified foremost by the diameter of its entrance aperture, which is usually the clear diameter of the first optical element, e.g., the objective lens or primary mirror. This largely dictates the light-collecting area of the telescope and its ability to resolve fine details in the image, both of which are very important considerations for astronomical telescopes. Considerable effort and expense are applied to the construction and support of the primary optics and any others (e.g. secondary mirrors) required to produce the first, real image of a distant object. Although the eyepiece is generally a much smaller and typically less expensive optic, it possesses much greater optical power than the primary telescope optics. We defer defining precisely what we mean by optical power until Section 2.3, but for now, it can be thought of as the ability of a lens or system of lenses to deviate light. It is because of the greater optical power of the eyepiece that the overall image quality in the telescope owes much to its design, and why high-performance eyepieces command a premium price. The principal purpose of this book is therefore to explore the choices and compromises that influence the design and affect the image quality of tele-scope eyepieces, predominantly for astronomical usage.

An important optical difference between an eyepiece and a magni-fying glass (or other magnifiers) is that magnifiers are typically used in isolation from other optics, whereas an eyepiece is used in conjunction with additional optics, viz. the main optics of the telescope. The entrance aperture of the telescope, which determines what light is admitted to the system, also serves as an important optical reference plane called the aperture stop. Two associated planes, the exit pupil of the main telescope optics and the exit pupil of the telescope+eyepiece combination, also need to be considered and will be defined later. Any rays passing through the centre of the aperture stop are referred to as principal rays or chief rays, whereas rays passing through the edge of the aperture stop are called marginal rays; we shall make use of this ray terminology extensively in this book. There are many other rays filling the aperture between these two extremes. Although the aperture stop of the telescope optical system is established by the telescope, we will find that it sets an important con-straint on eyepiece design.

Another constraint comes from the necessity of the eyepiece to produce images at much larger field angles, in the range 20°–55° to suit the human eye, compared to just 1° or so for the main telescope optics.

The primary optics, whatever their detailed configuration, deliver to the eyepiece a number of cones of light, one per object point. The cone of rays originating from a single object point is referred to as a pencil; think of the conical, tapered end of a pencil, which then changes to a parallel-sided shape, and you can see the association with the shape of converging, diverging and collimated ray paths. The rays within each pencil converge in the telescope focal plane, forming a real image of the associated object point, and then diverge again, until they are intercepted by the optics of the eyepiece. Some eyepiece types intercept the rays *before* they reach the telescope focal plane; that arrangement doesn't particularly complicate matters and, as we will see in Chapter 5, can assist. The convergence angle of a cone of rays depends jointly on the diameter of the entrance aperture D of the telescope and the effective focal length f' of its main optics. The convergence angle could be specified in degrees, but it is more common in the context of telescopes and eyepieces (and cameras, as it happens, but not microscopes) to describe it in terms of the ratio f'/D, stated in the format "f/ratio". For example, a telescope with effective focal length $f' = 3000\,\text{mm}$ and aperture $D = 300\,\text{mm}$, whose f-ratio is therefore $f'/D = 3000/300 = 10$, would be stated to be "an $f/10$ system", meaning the diameter is 1/10th of the effective focal length, quite literally $D = f'/10$.[1]

Although I have emphasised several times already that this book is principally about eyepieces used in telescopes, astronomical telescopes in particular, it is nevertheless the case that eyepieces for other optical instruments share many of their features, and therefore it is hoped this book will be of value to people using other similar instruments such as microscopes and terrestrial telescopes. This book could also form the basis of an introductory course in optics, optical aberrations and optical design, bridging the gap between introductory books and advanced optical-design texts.

1.2 LAYOUT OF THIS BOOK

This book is divided into six chapters. Chapter 2 covers the optics knowledge needed to understand the basic features of eyepiece designs. In Chapter 3, we examine a trivial, single-lens eyepiece and see how the many deficiencies in its optical design – so-called aberrations – arise and how they may be quantified, visualised and evaluated. While a single-lens eyepiece would almost never deliver adequate image quality to justify its use, examining it is a useful exercise as a precursor to studying better, more complex optical designs. The final section of Chapter 3 covers a mathematical approach to wavefront analysis – Zernike analysis – that we do not make further use of in this book, and this section could therefore be skipped without detracting from your understanding of the subsequent material; it is there for completeness.

Chapters 4 and 5 examine a broad range of eyepieces utilising spherical optical surfaces. These account for the vast majority of real eyepieces in use both historically and at present. The content of these chapters is sequenced in a way that traces the logical development of eyepiece design, and therefore it is partly chronological, but not rigidly so. In particular, Chapter 4 covers the historical two-lens eyepieces of the Huygens and Ramsden designs, while Chapter 5 covers most subsequent forms (mostly 19th and 20th century) that have progressively improved eyepiece performance.

Across all of Chapters 3–5, much use is made of modern computer-based methods of raytracing to investigate and illustrate the aberrations of eyepieces, and to see how control of the design changes the inherent aberrations. The "WinLens Basic" ray-tracing programme (V1.2.11; 2019-04), written by Dr Geoff Adams of OpticalSoftware.net and distributed by Qioptiq®, has been available free for many years, operating under Windows®, and it is used for these comparisons. A sufficiently motivated reader may wish to install the programme for their own use to replicate and extend some of the comparisons made in this book.

Eyepiece design remains an active endeavour, and changing optical and manufacturing technologies are allowing more advanced eyepieces to be envisaged. Chapter 6 considers more advanced eyepiece designs or influences, including antireflection coatings, aspherical surfaces and accessories.

Much has been written online about astronomical eyepieces, but it can be difficult to ensure accuracy or to find sufficient detail, and inconsistencies are not uncommon. I have therefore endeavoured to trace information back to primary sources, particularly patents and manufacturers' specifications or to other professionally edited textbooks where possible, all of which are referenced in extensive endnotes to each chapter.

1.3 MATHEMATICS IN OPTICS

Previous books containing chapters or significant sections on eyepieces have typically fallen into two camps: those which concentrate mainly on the qualitative features of the images they deliver in a largely non-mathematical way,[2] and those written for expert optical physicists and designers[3,4,5] which assume the reader already has an advanced conceptual grounding in optics and can perform Olympic-standard mathematical gymnastics.

The aim of this book is to find the middle ground: to provide sufficient mathematical background in optics to enable the reader to understand *why* certain aspects of optics may aid or frustrate the desire to produce a good image from a telescope eyepiece, and also sufficient explanation to help the reader *interpret* the results and make sense of the visual outcomes. This book avoids topics in optics that are not essential to understanding eyepieces, such as telescope mirrors, to keep the narrative focussed.

Most of the required mathematics is either algebra or trigonometry, the latter forming the basis of many ray diagrams, which can usually be broken down into triangles. A hurried reader who is not presently in the frame of mind to follow a mathematical derivation could skim over some material, seeking only to note the equations derived and the definitions of the variables within them, but skipping mathematical material entirely would inevitably limit a reader's understanding of the subject, so I would encourage persevering with the algebra to grasp as much of this subject as you can. Most rules of algebra reduce to being able to add, subtract, multiply and divide in the correct order, and most trigonometry reduces to knowledge of the sine function (see below). A Glossary of Variables and Points is provided as a quick reminder of the meaning of mathematical symbols used in this book.

To aid the reader, diagrams have been carefully drawn to indicate visually the relationships between various optical parameters, and a consistent approach has been taken to formatting the symbols in equations and diagrams. Variables, i.e., symbols which stand for a numerical value and its units, are shown in italic font (such as n, s and α) while labels, which denote points in space or some other non-numerical information, are shown in non-italic font (such as V, C and S). Mathematical symbols are of course a shorthand form of conveying information, so small differences between otherwise similar symbols are important. As examples, a lower-case italic s and an uppercase non-italic S convey different information, as do the five distinct symbols f, f', F, F and F$'$, which will be introduced later. It won't surprise you to learn that these five "f"s all relate to the focussing performance of an optical surface or lens, but they tell us different things about that performance, and hence they are distinct. What matters most is that when you see them, you *think in words about the optical quantity they are intended to represent*, not merely what letter of the alphabet they most resemble, so for example read f' as "second focal length", not "eff-prime". The Glossary of Variables and Points is located at the start of this book, preceding Chapter 1.

(a)

h

o

α

a

$\sin \alpha \equiv o/h$
$\cos \alpha \equiv a/h$
$\tan \alpha \equiv o/a$

(b)

a C

B b

c A

$(\sin A)/a = (\sin B)/b = (\sin C)/c$

(c)

h

θ

$o \approx h\theta$

$a \approx h$

$\sin \theta \approx h\theta/h = \theta$
$\cos \theta \approx h/h = 1$
$\tan \alpha \approx h\theta/h = \theta$

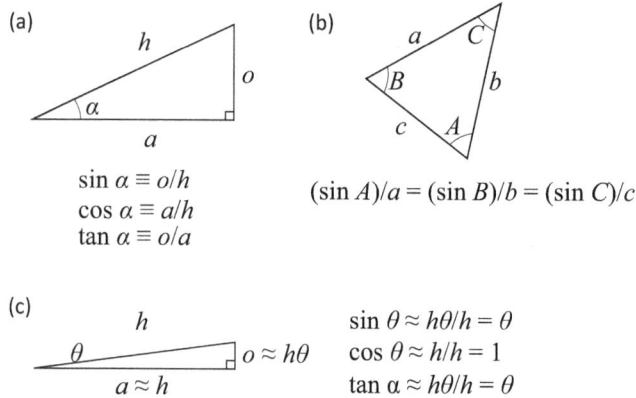

FIGURE 1.1   Sine, cosine and tangent functions and sine rule (a) The sine of an angle α is defined for a right-angle triangle as the ratio of lengths of the opposite and hypotenuse sides, $\sin \alpha \equiv o/h$. The cosine of an angle α is defined for a right-angle triangle as the ratio of lengths of the adjacent and hypotenuse sides, $\cos \alpha \equiv a/h$. The tangent of an angle α is defined for a right-angle triangle as the ratio of the lengths of the opposite and adjacent sides, $\tan \alpha \equiv o/a$. (b) The sine rule gives the relationship between the side lengths and the opposing angles in a triangle of arbitrary shape. (c) If the angle of interest is very small, then the trigonometric functions can be approximated somewhat crudely as $\sin \theta = \theta$, $\cos \theta = 1$ and $\tan \theta = \theta$, provided the angle θ is expressed in radians, not degrees.

As a quick reminder of some relevant trigonometry, Figure 1.1a indicates the definition of the sine (sin), cosine (cos) and tangent (tan) function in a right-angled triangle. Figure 1.1b conveys the sine rule, which relates side lengths and angles in a general triangle. If the angle of interest is very small, so the triangle is long and thin as in Figure 1.1c, then the trigonometric functions can be approximated as $\sin \theta \approx \theta$, $\cos \theta \approx 1$ and $\tan \theta \approx \theta$ provided the angle is expressed in radians, which is an alternative angular measure to the degree.[6] We will often use these and some more accurate approximations in subsequent chapters to simplify the analysis of ray paths.

NOTES

1 For an on-axis object, the angle between the marginal rays of a converging cone and the optical axis, which we can call the semi-angle α of the cone, will have a tangent (see Figure 1.1a) given by $\tan \alpha = (D/2)/f' = 1/(2 \times f\text{-ratio})$. Another way of describing the converging cone is via a quantity called the numerical aperture NA, defined as $NA \equiv n' \sin \alpha$, where n' is the refractive index of the medium in which the cone is converging (typically air, in

which case $n' = 1.00$), and α is again the semi-angle of the cone. The numerical aperture is commonly quoted in microscopy, whereas in astronomy and photography f/ratios are more commonly quoted. For small cone angles, $\sin \alpha \approx \tan \alpha$ and then $NA \approx 0.5/f$-ratio.

2 H. Rutten and M. van Venrooij, *Telescope Optics Evaluation and Design*, Willmann-Bell Inc., 1988, Chapter 16.

3 M. Born and E. Wolf, *Principle of Optics*, 7th edition, Cambridge University Press, 1999.

4 M.J. Kidger, *Fundamental Optical Design*, SPIE, 2000, Chapter 11.

5 R. Kingslake and R.B. Johnson, *Lens Design Fundamentals*, 2nd edition, Academic Press, 2010.

6 A circle is commonly divided into 360 uniform intervals of angle called degrees, but it can alternatively be divided into 2π uniform intervals called radians. Therefore 1 radian $= 360°/2\pi \approx 360°/6 = 60°$. Both angular measures are useful in optics and will be used wherever appropriate.

CHAPTER 2

Optics Fundamentals

2.1 REFRACTION

Light falling on a polished, glass-like material is partially reflected and partially transmitted. Unless the light arrives perpendicular to the surface, the transmitted light is refracted, i.e., its direction of travel is changed, and it also experiences absorption and scattering in the glass, though a good quality optical glass will minimise those fates. Understanding refraction is key to eyepiece design. Partial reflection is also important in eyepieces, and becomes problematic if not adequately addressed, a point we shall return to in Section 6.1.

The refraction of light is often illustrated using ray diagrams, which show, for example, how the path of a ray of light is deviated upon entering or leaving a piece of glass. However, rays are only a conceptual convenience used in elementary attempts to describe the passage of light; light is not made of rays and certainly doesn't travel in straight lines. While we shall certainly speak of rays extensively in this book and use them to describe the path of light in many situations, we ought to have a more realistic understanding of what light is.

Light propagates as oscillating electric and magnetic fields, i.e. as waves, with the distance between successive crests of the oscillation being called the wavelength. For visible light, i.e., light that excites the photoreceptors in the human eye, the wavelength is typically in the range 0.4–0.7 μm (micrometres or commonly microns), or equivalently 400–700 nm (nanometres), where 1 micron is 1/1000th of a millimetre, and 1 nm is 1/1000th of a micron. When light waves propagate outward from a source, they do so in all directions, unless the paths are restricted by materials that

DOI: 10.1201/9781003670506-2

(a)

(b)

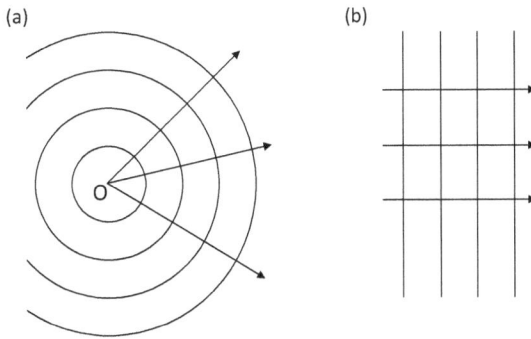

FIGURE 2.1 Wavefronts and rays. (a) Light emitted from a point source O consists of oscillating electric and magnetic fields that propagate outwards at a high speed in all directions. Wave crests emitted at the same instant constitute an expanding, 3D, spherical wavefront, shown on this flat page in 2D as expanding circles. It can sometimes be helpful to draw lines showing the directions in which the wavefront is moving; such lines are called rays (here illustrated by arrows). (b) At a very large distance from a point source, a small portion of a large spherical wavefront may be treated as a series of flat wavefronts, here moving from left to right. For flat wavefronts, the rays (arrows) will be parallel to one another.

absorb, reflect or refract the waves. The crests of the wave emitted from a particular source at a given instant are described as a wavefront and, if unimpeded, spread out in all directions from a point source as a spherical shell (Figure 2.1a). What we speak of as rays are simply arrows that tell us the direction the wavefront is moving, and the case shown in Figure 2.1a would be described as a divergent pencil of rays. In optics, a "pencil" refers to rays originating from the same source point or object point. At a very great distance from the source, the small portion of a spherical wavefront that we may encounter will appear as a flat wavefront; using the naïve ray description, we would say the rays are then moving parallel to one another (Figure 2.1b). It is easy to make the mistake of thinking the rays "are" the light; rays are useful for simplifying calculations and visualising how light travels, but sometimes they oversimplify the picture by neglecting important properties of light waves.

The processes we recognise as reflection and refraction are due to a redirection of the wavefront, and both are a consequence of the wave nature of light. Reflection arises when there is a change in the direction of wave propagation without the waves moving into a new medium, such as when waves travelling through air strike the surface of a metallic mirror and are reflected, still travelling in air. However, if the wavefront moves from

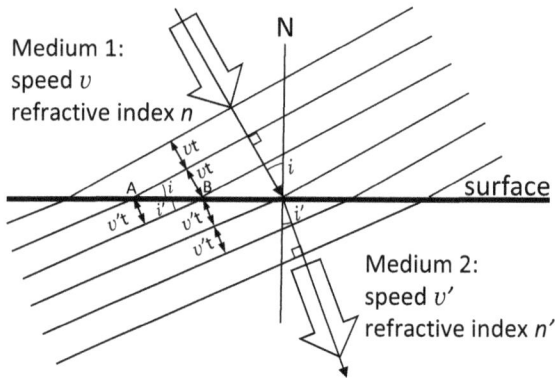

FIGURE 2.2 Refraction of plane waves by a flat optical surface. A series of flat wavefronts arrives at a flat surface (bold), which is the boundary between two different media, where the propagation speed decreases from v to v'. This could be an air-glass surface, or the surface between two types of glass, for example. As a result of the decreased speed in the second medium, wavefronts that previously travelled a distance vt in the first medium over a time t now travel a smaller distance $v't$ in the second medium. The reduced speed retards the progression of the wavefront, effectively bending it and thus changing its direction of propagation.

one medium to another across a surface, such as from air into glass or vice versa, the speed of wave propagation changes. This also changes the *direction* of propagation if the wavefront is inclined to the surface, because different points along the wavefront arrive at and progress beyond the surface at slightly different times, and travel at different speeds. This case is illustrated in Figure 2.2 for a flat wavefront encountering a flat air-glass surface, where the incoming wavefront is inclined to the surface by some angle. The result of the reduction in speed upon entering the glass is that the propagation distance within a given time interval decreases, and this is manifested as a tilt in the wavefront. If we revert to the ray explanation, we would say that the rays have, for some reason, changed direction in Figure 2.2, but this only occurs as a consequence of the wavefront being tilted by the change in speed; rays divorced from wavefronts would have no reason to bend. Nevertheless, as a means of showing the direction the wavefront is moving, rays provide a convenient illustration of the passage of light, and we can even apply a mathematical analysis to their progress to determine much, though not all, about how a wavefront moves in an optical system. We shall therefore continue to draw diagrams of rays where it suits us to do so, but bearing in mind that they provide a limited treatment of what light is and how it behaves. Occasionally, we have to revert to a proper wavefront analysis to understand how light behaves in an optical system.

Law of Refraction

The analysis of wavefront refraction in Figure 2.2 shows that the change in the direction of propagation of the light depends on its speed in the two media.[1] When the wavefront arrives, its direction of propagation is at an angle i relative to the normal N to the surface. (The "normal" is the direction at right angles to the surface.) The incoming wavefronts make the same angle i with the surface, and considering the small line segment length $\overline{AB}$ at the surface, we see that $\sin i = vt/\overline{AB}$.

Considering the outgoing wavefronts, their direction of propagation makes an angle i' with the normal N, and their wavefronts are inclined by the angle i' relative to the surface, so considering again the surface line segment length $\overline{AB}$, we can write $\sin i' = v'/\overline{AB}$.

Rearranging both equations above as expressions for $1/\overline{AB}$, and recognising that both forms must be equal, gives us $\dfrac{\sin i}{vt} = \dfrac{\sin i'}{v't}$. The common factor of t can be multiplied out, then multiplying both sides by the speed of light in vacuum, c, gives $(c/v) \sin i = (c/v') \sin i'$. It is customary to define the speed ratio c/v as a quantity called the refractive index, n, where $n \equiv c/v$ and typically takes values around 1.50–1.80 for glass, while for air it is ~1.0003, close to 1.00. The "$\equiv$" sign as used here indicates that the refractive index is *defined* to be the ratio c/v and thus always equals this; when we read "n" and say "refractive index", it is our shorthand way of saying and thinking "the speed ratio c/v". Similarly, $n' \equiv c/v'$. Consequently, the directions of propagation i and i' are related to the refractive indices (speed ratios) n and n' on the two sides of the surface by the equation

$$n \sin i = n' \sin i' \qquad (2.1)$$

which has come to be known as Snell's law.

Imagine we trace a ray of light from some point O to a second point O'. A very useful feature of light is that if we reverse the direction of the ray at point O', it will retrace the same path back to point O. This feature is called the reversibility of light and has its origins in a more fundamental piece of physics called Fermat's Principle; the details of that need not concern us here but make for interesting further reading. One practical implication is that while we may wish to design a series of lenses to convey light from point O to point O', we can actually design it back to front, i.e., to convey light from point O' to point O, and it will perform the same task. As you will see later in the book, eyepieces are nowadays almost always designed reversed, though the earliest designs we will encounter were not.

The reversibility of light also means that the terms "angle of incidence" and "angle of refraction", which are commonly used in introductory optics books, are slight misnomers because they may mistakenly imply that the path depends on the direction the light is travelling, i.e., on which side of the optical surface the light is incident, and on which side it is refracted, but the reversibility of light means the distinction is moot; the result would be the same if we traced the ray the other way. So, in the form of Snell's law provided in Equation 2.1, the convention has been adopted to use variables without primes, i.e., n and i, on one side of the refracting surface, and to use variables with primes, n' and i', on the other side of the surface, without artificially distinguishing between which is the incident side and which is the refracted side. In practice, as we trace rays through optical surfaces, we will usually use non-primed variables for the object side and primed variables for the image side, but mathematically the distinction is tenuous; if you swap over the primed and unprimed variables in Equation 2.1, you still end up with the same equation. Moreover, we will shortly blur the distinction between object and image, and refer to them collectively as conjugates of one another.

The analyses of light using a wave treatment or a ray treatment are termed physical optics and geometric optics, respectively. Geometric optics uses the ray approximation to analyse the passage of light through an optical system, but occasionally the wave treatment is required to get an accurate answer. A good example of the latter would be the necessity of using wave optics to calculate the diffraction limit of a telescope to determine its resolution. However, geometric optics will suffice for most of our analyses of eyepieces.

2.2 DISPERSION

As noted in Section 2.1, the speed of light decreases when it passes from air into glass, and the speed ratio relative to that in vacuum is defined as the refractive index, $n \equiv c/v$. It is important to note that different materials have different refractive indices, and furthermore, they depend on wavelength, as the speed of propagation depends on wavelength. At visible wavelengths, optical materials generally have a higher refractive index at shorter wavelengths, and the variation with wavelength also increases towards shorter wavelengths. Refractive indices of four glass types are shown in Figure 2.3a. The retinal cone cells that render the human eye sensitive to colour[2] are sensitive to a wavelength range from around 400 nm (violet) to 700 nm (red), which the data range in the figure has been set to match.

(a)

(b)

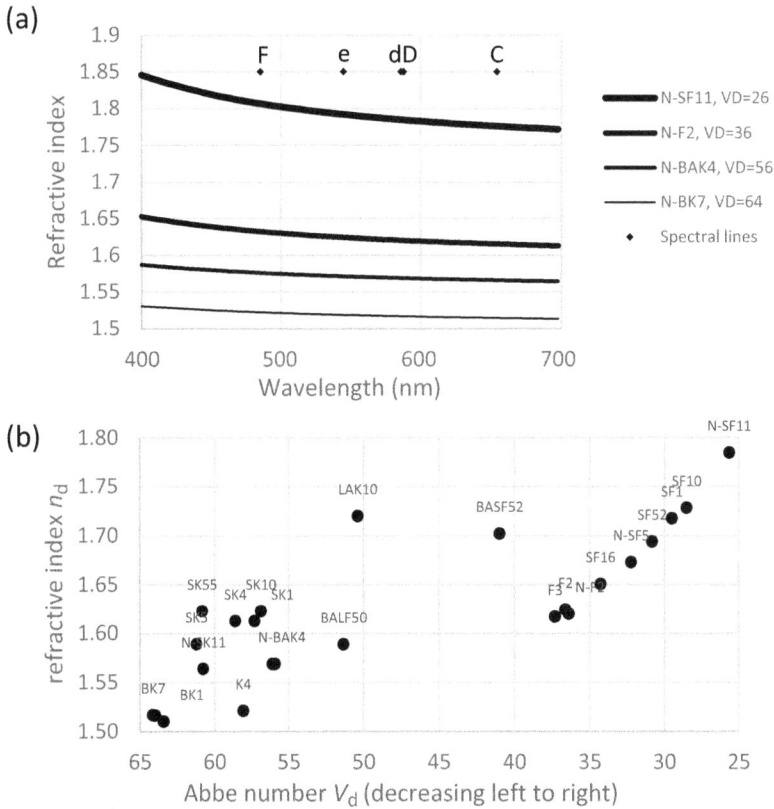

FIGURE 2.3 Refractive index of optical glass. (a) Refractive index varies with wavelength. Refractive indices of four optical glasses from the Schott glass catalogue are shown: borosilicate crown glass N-BK7,[3] barium crown glass N-BAK4,[4] flint glass N-F2[5], and dense flint N-SF11[6]. At visible wavelengths (400–700 nm), refractive indices decline towards longer wavelengths. Borosilicate crown (BK) glass has the addition of boric oxide to an alkali lime silicate glass, while barium crowns (BaK or BAK) have barium oxide added. Higher density additives such as titanium dioxide or zirconium dioxide (which now replace the traditional lead oxide) make flint (F) and dense flint (SF) glasses which have even higher refractive indices. The "N-"prefix indicates these are new (modern) variants with similar optical properties to the original variants, but with modified compositions. Five common reference wavelengths are marked according to their Fraunhofer designations F, e, d, D and C, as described in the text. (b) Refractive index and dispersive power (Abbe V-number) of different glasses as labelled. These differences lend themselves to designers pairing glasses having complementary properties, allowing light of different wavelengths to be better managed in the optical design process (see Section 3.8).

It is obvious from Figure 2.3a that the difference in refractive index between blue (~450 nm) and red (~600 nm) wavelengths is greater for the two flint glasses than for the two crown glasses. Because differences in refractive index translate into different angles of refraction, red and blue pencils of light from the same source point separate as they are refracted. This behaviour is called dispersion. The standard way to quantify it is to calculate the difference in refractive index between blue light at 486.1 nm and red light at 656.3 nm, n_{486}-n_{656}, and to divide this into the quantity n_{589}-1, where n_{589} is the refractive index of orange light at 589.3 nm. These three wavelengths are more commonly known by the letter designations given by Fraunhofer[7] to the spectral lines at these wavelengths: F, C and D, respectively (Figure 2.3a). The resulting ratio is known variously as the dispersive index,[8] Abbe number, V value or constringence,[9] and is given formally as

$$V_D \equiv \frac{n_D - 1}{n_F - n_C} \qquad (2.2)$$

The numerator, $n_D - 1$, is typically around 0.5 for crown glasses, ranging up to around 0.8 for dense flints. Alternative definitions often replace the D line (Na 589.3 nm) with the d line (He 587.6 nm) or less commonly with the e line (Hg 546.1 nm)[10]; the last of these is potentially preferred for lenses intended for visual observing[11] since the peak of sensitivity of human rod cells is around 500 nm, while S, M and L cone cells peak at 420, 534 and 563 nm, respectively,[12] making the e line the most similar to the peak of human vision. Nevertheless, the Abbe number calculation is dominated by the denominator; a greater difference between the blue and red refractive indices increases the denominator and hence gives a smaller Abbe number. The Abbe values for the four glasses shown in Figure 2.3a are given in the legend and decrease from high values such as 64 and 56 in N-BK7 and N-BAK4 (crown glasses) to low values such as 36 and 26 in N-F2 and N-SF11 (flint glasses). Values for a wider range of glass types are shown in Figure 2.3b; note that the horizontal axis *decreases* towards the right, corresponding to more dispersive glass having lower V-numbers. The particular glass LaK10, a lanthanum crown, is notable as having a relatively high refractive index but low dispersion (high V-value). The potential advantage of this and similar glasses will become clearer in Section 2.3 and Chapter 3.

The *reason* for the increase in refractive index with decreasing wavelength is linked to the fact that these materials have been selected as highly transparent in visible light, but they are strongly absorbing at slightly shorter wavelengths around 200–300 nm, i.e., in the ultraviolet. Stronger absorption of light occurs when the frequency of oscillation of the light wave approaches a natural resonant frequency of the molecules it encounters. The electric field oscillations of the light induce oscillations of the charges in the material of the glass, which in turn generate secondary light waves. Those secondary waves travel predominantly in the same forward direction as the incident waves and combine with them, though with a small phase difference, which manifests as a change in wave speed, and we register that as a change in refractive index. At wavelengths closer to the peak absorption wavelength, the changes in speed and therefore refractive index become more striking.[13] As a corollary, the datasheet for N-BK7[14] shows its transmission falls from 92% at 400 nm to just 5% at 285 nm (for a 10 mm thick sample), while that for N-SF11 falls from 46% at 400 nm to 5% at 375 nm, i.e. over just 25 nm.

The dependence of refractive index on wavelength inevitably poses a challenge to eyepiece designers who must therefore attempt to counteract the dispersion introduced by one lens through an opposing dispersion produced in another lens. This balancing act is one of many that an optical designer faces, as will become more apparent as the design story of this book unfolds, especially in Chapter 3.

2.3 REFRACTION AT SPHERICAL OPTICAL SURFACES

Anyone who has admired a telescope, microscope, camera or eyepiece could be forgiven for thinking that the basic design element of an optical system is a lens. After all, isn't that what lens designers do, design lenses? However, it is much more useful to regard the basic design element of optics to be a surface. A lens is "merely" two optical surfaces stuck together with glass; refraction only takes place at the two surfaces, where the refractive index changes.

An optical surface is described by its shape and the refractive indices of the media through which light travels immediately before and after encountering the surface. In the case of a front-surface mirror, there is no change in medium for the incident and reflected rays, so the surface shape is the sole determinant of the change in the light path. In contrast, for a refracting surface, Snell's law (Equation 2.1) indicates that the angle at which the refracted ray leaves the surface (i') depends on the two refractive indices (n, n') and the angle at which the incident ray arrives at the surface (i), recalling that both i and i' are measured *relative to the normal to the surface*. On a curved refracting surface, the direction that the normal makes relative to the incident light varies from point to point across the surface, so we need a way of calculating these variations. To do this, we have to describe the shape of the surface mathematically. The simplest curved surface is spherical, as every point on the surface lies a fixed distance – the radius of curvature – from a single point called the centre of curvature. Other surface shapes are possible, and we examine non-spherical surfaces in Section 6.3, but as spherical surfaces are also amongst the easiest and least expensive to produce by traditional glass-grinding and polishing techniques, they make a useful starting point.

The analysis that follows is based on the layout in Figure 2.4, which considers two rays from a point source O located a specified distance s from the vertex V of a spherical surface. Like most ray-tracing studies, we assume light travels from left to right (in the absence of reflections). The surface is fully described by its radius of curvature r and the two refractive indices n and n' that define the boundary. The line joining the object point O and the centre of curvature C of the surface is called the optical axis. The ray launched along the optical axis reaches the vertex normal to the surface at that point and hence undergoes no refraction; it passes on along the optical axis. However, the second ray launched at an angle α to the optical axis

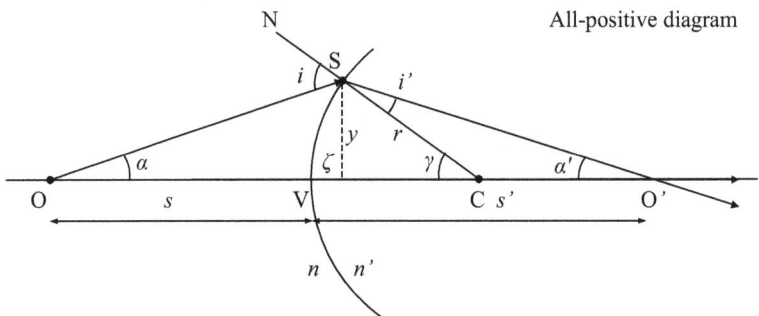

FIGURE 2.4 Ray-tracing through a spherical surface (all-positive diagram). Two rays are launched from an object point O, located a distance s in front of a spherical refracting surface having radius of curvature r: one ray passes horizontally along the optical axis, passing through the surface at its vertex V, while the second ray is launched at an angle α to the optical axis. The refractive indices of the medium to the left and right of the surface are n and n', respectively. All other quantities, such as the location of the point S at which the inclined ray intercepts the surface, the distance s' at which the two transmitted rays intersect again forming an image O' of the object, and the angle α' at which the refracted ray converges with the optical axis, can be calculated exactly. The resulting ray-tracing formulae are derived in the text.

reaches the surface at the point S, and is clearly inclined relative to the normal (N) to the surface at that point, having an angle of incidence i. It therefore undergoes refraction. This particular diagram is drawn for the case where $n' > n$, so the ray is refracted downwards and intersects the first ray at the point O'. The point O' therefore corresponds to the image of point O, as two rays diverging from O have converged again at O'. Other rays launched from O at the same angle α to the optical axis, such as those inclined downwards or into or out of the page at that same angle, will also converge at O'. Ideally, rays launched at other angles $\neq\alpha$ would also converge at O', and the image formed at O' will be of high quality; we shall soon see the conditions under which this is true enough, and later (Chapter 3) we will examine the exceptions in detail, as that is how we can identify and correct imaging errors (aberrations) and render high quality images.

Our challenge initially is to obtain mathematical expressions for three quantities:

- the location S at which the inclined ray encounters the surface

- the distance s' from the vertex V to the image O'

- the angle α' at which the refracted ray crosses the optical axis at O'

We can achieve this, exactly, using trigonometry.

Ray Tracing an All-Positive Spherical Surface

We begin by considering the triangle OSC in Figure 2.4. The side CS is extended as a straight line to delineate the normal N to the surface at S. The ray along OS makes an angle of incidence i at S, so angle OSC is evidently $180° - i$. We use the sine rule (Figure 1.1) to write $\dfrac{\sin(180° - i)}{s+r} = \dfrac{\sin\alpha}{r}$, but since $\sin(180° - i) = \sin i$, we can recast this as $\sin i = \dfrac{s+r}{r}\sin\alpha$ where all quantities on the right-hand side are known, and thus the angle of incidence i can be deduced.

The refractive indices on both sides of the surface are also known, so Snell's law (Equation 2.1) quickly yields the angle of refraction i' via $\sin i' = \dfrac{n}{n'}\sin i$.

The internal angles of the triangle OSC must sum to 180°. We have labelled the angle SCO as γ, so $\gamma + \alpha + (180° - i) = 180°$, hence we can calculate $\gamma = i - \alpha$. We can therefore deduce that point S lies at a height

$y = r \sin \gamma$ above the optical axis, and a horizontal distance $\zeta = r - r \cos \gamma$ behind the vertex V, so we have now located the point S where the ray encounters the surface via its coordinates (ζ, y).

It remains only to calculate the location of the point O' at which the image of O is formed, and the angle α' at which the refracted ray meets the optical axis. We turn our attention first to the triangle OSO', whose interior angles must add to 180°, so we can write $\alpha + [(180° - i) + i'] + \alpha' = 180°$, and therefore deduce that $\alpha' = i - i' - \alpha$. Finally, we apply the sine rule to the triangle SCO' to write $\dfrac{\sin i'}{s' - r} = \dfrac{\sin \alpha'}{r}$, and thus we can write $s' = r\left(1 + \dfrac{\sin i'}{\sin \alpha'}\right)$.

To recap, we have started from a spherical surface specification r, n and n', and launched a pair of rays from a point O located a distance s from the surface vertex V. One of these rays travels along the optical axis while the second is inclined at an angle α to the optical axis, and we have deduced that the two rays must converge to form an image of O at a distance s' behind the vertex, with the refracted ray converging at an angle α'. These calculations are exact and allow us to trace a plethora of rays at different launch angles α, to determine how close to one another the various rays intersect the optical axis, i.e., whether the image O' is good or poor.

The ray-tracing steps outlined above provide a reasonably straightforward explanation of how ray paths can be calculated exactly for a spherical surface. However, this is not necessarily the best way of proceeding long-term, for reasons we must now explore. Firstly, the approach was derived based on Figure 2.4, which was referred to as an "all positive" diagram. This means that in the diagram and the subsequent analysis, all variables conveniently took positive values. This convenience soon vanishes if the passage of the rays through subsequent surfaces is also to be calculated, and it is relatively easy to see why. The incident ray along OS and the refracted ray along SO′ make positive angles α and α' with the optical axis, but it is clear that the ray along OS is ascending whereas the ray along SO′ is descending, and this difference must be recognisable in the calculations. Furthermore, if we choose a concave optical surface curving to the left instead of the convex surface curving to the right, the diagram and equations would have to be modified, which could be inconvenient. Finally, we may encounter cases where the refracted rays are directed above the horizontal and thus away from the optical axis, rather than towards it, and for which there is no real point of convergence; in this case the point O' is a virtual intersection to the left of the surface, identifiable only by projecting the refracted rays backwards to see where they appear to have come from, and in this case the surface would produce a virtual image of O.

The problems identified in the preceding paragraph are resolved by paying closer attention to the definitions of the signs of the quantities used in the calculations, i.e., whether they are positive or negative. Several different sign conventions exist in optics, so you must be careful not to combine the approaches from several different books if you haven't already ensured they adopt the same definitions. In this book, I will adopt a Cartesian system[15,16,17] in preference to the more baroque system adopted in many (especially older) physics texts.[18,19] A Cartesian system is also more likely to be utilised in computer-based ray-tracing software.

In implementing a Cartesian system, we place the origin of the coordinate system at the vertex of the surface, V, so distances to the left will henceforth be treated as negative, while distances measured to the right will be treated as positive. The adjustments we then must make to the equations derived above are to use the variables l and l' to refer to the distances of O and O' from V, instead of s and s'. Since a positive value of s will be replaced with a negative value of l, we make the replacement $s = -l$ in the equations. If the centre of curvature of the surface, C, is to the right of the vertex, then the radius of curvature r is treated as positive, but r will have a negative value for a concave surface, where the centre of curvature is to the left of the vertex. A negative value for r will also imply negative values for i and i' (corresponding to the ascending incident ray approaching the surface above the normal in this configuration), and these in turn will imply negative values for y, ζ and s', but all

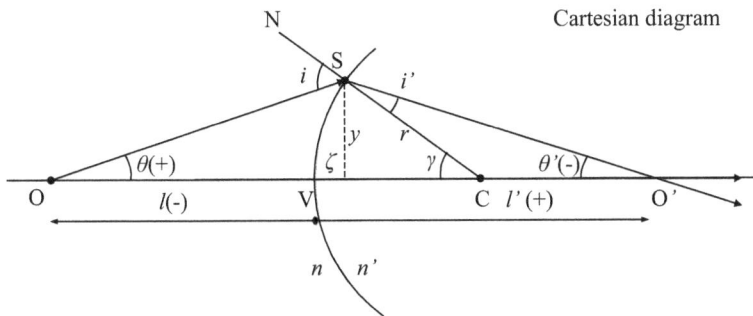

FIGURE 2.5 Ray-tracing through a spherical surface (Cartesian diagram). Similar to Figure 2.4, but using a Cartesian coordinate system with the origin at the vertex V of the surface. Distances measured to the left of the vertex take a negative value, and rays intersecting the optical axis with a downward trajectory are assigned a negative angle. The signs appearing in brackets in the figure emphasise where positive and negative values may commonly arise. Distances below the optical axis also have negative values.

of these impacts are appropriate in a Cartesian system. The final change we need to implement is to define the signs for the angles that the rays make with the optical axis. Even amongst other users of Cartesian conventions, the ray angle slope can be defined in opposite ways[20,21]! In this work I adopt the approach[22] that the Cartesian slope of the ray along OS (moving upwards as it travels from left to right) should be regarded as positive, so α retains its positive sign, but the slope of the refracted ray along SO′ is then negative so the convergence angle α' must switch to a negative value. This is achieved if we switch the angle notation from α and α' to θ and θ', with the replacement $\alpha' = -\theta'$. Conveniently, this definition is also consistent with the sign for i and i' being negative in the concave case, where the ray converges with the normal from above, mimicking θ' being negative if the ray converges with the optical axis from above.

The ray-tracing formulae derived above, adjusted to the Cartesian reference system depicted in Figure 2.5, can be summarised thus:

$$\sin i = \frac{r-l}{r}\sin\theta \tag{2.3a}$$

$$\sin i' = \frac{n}{n'}\sin i \tag{2.3b}$$

$$\gamma = i - \theta \tag{2.3c}$$

$$y = r\sin\gamma \tag{2.3d}$$

$$\zeta = r - r\cos\gamma \tag{2.3e}$$

$$\theta' = i' - i + \theta = i' - \gamma \tag{2.3f}$$

$$l' = r\left(1 - \frac{\sin i'}{\sin\theta'}\right) \tag{2.3g}$$

Alternative approaches, even within the Cartesian convention, can be established[23]; there is often more than one way to solve a mathematical puzzle.

For completeness, we note that if O is placed a very long way to the left, essentially at infinity, then the incident angle $\theta = 0$ and the first equation in the sequence of Equations 2.3a-g will not be able to initiate a sensible ray trace. In this case, we must instead specify a realistic ray height y rather than angle θ, use y to calculate γ by solving $y = r \sin \gamma$, and then calculate i (which in this case equals γ). The calculations of i', ζ, θ' and l' can then proceed as before.

The ray-tracing steps outlined above provide a computational means of determining where a spherical refracting surface will direct one or more incident rays, but it does not provide much intuitive insight into how or why the image is formed where it is. To achieve a better level of *intuitive* insight, it is helpful to simplify the problem by restricting the analysis to the subset of rays that lie close to the optical axis. This approach is called variously the paraxial or Gaussian approximation (after its originator, K.F. Gauss, who set out the theory in 1841), and is equivalent to saying we will only consider small values of θ, which implies small values of i, i' and θ'.

Small-Angle Approximation for the Sine Function

Snell's law (Figure 2.2 and Equation 2.1) shows that the tilting of the wave-fronts at an optical surface depends on the sine of the angles of incidence and refraction. Angles can be specified either in degrees or radians, and although degrees are used more in everyday life, radians provide advantages mathematically.[24] As a case in point, an important feature of the sine function is that it can be written as a series: if the angle of interest (x, say) is expressed in radians, then[25]

$$\sin(x) = x - x^3/3! + x^5/5! - x^7/7! + \cdots \qquad (2.4)$$

If x is less than about 0.5 radians (corresponding to $\approx 30°$), the successive terms become very small. For example, if we evaluate this series for $x = 0.5$, we find

$$\sin(0.5) = 0.5000 - 0.0208 + 0.0003 \cdots \approx 0.4795, \text{ cf. } \sin(0.5) = 0.4794\ldots$$

It is obvious that keeping only the first term introduces an error of only 4%, so making the approximation $\sin(x) \approx x$ (for x in radians, where $x < 0.5$) introduces only a small penalty in lost accuracy.

We can make use of the small-angle approximation $\sin(x) \approx x$ to remove the sin functions from the ray-tracing equations, both in Snell's law and elsewhere, obtaining a less precise but more intuitively informative set of equations to describe refraction by a spherical surface. This is called the paraxial approximation, and in the mathematics that follows, the approximation will be noted explicitly with the symbol "$=_p$" where the subscript "p" is inserted to remind us that the equation holds only for rays meeting the paraxial criterion set out above.

Paraxial Approximation for a Surface

If we start with Snell's law, $n \sin i = n' \sin i'$, and use the exact expressions derived above in Equation 2.3, that $i = \gamma + \theta$ and $i' = \gamma + \theta'$, we obtain $n \sin (\gamma + \theta) = n' \sin (\gamma + \theta')$. As we are considering only small angles in the paraxial approximation, we can strip out the sine functions and simplify this to $n(\gamma + \theta) =_p n'(\gamma + \theta')$, recalling that the angles must now be expressed in radians, not degrees.

Expanding the brackets and rearranging the terms, we obtain $-n'\theta' =_p - n\theta + \gamma(n' - n)$.

Turning back to the all-positive diagram in Figure 2.4, we can write $\sin \alpha = y / \overline{OS}$, $\sin \alpha' = y / \overline{SO'}$ and $\sin \gamma = y/r$. Imagining the paraxial case where α is very small so S is close to the vertex V, the lengths $\overline{OS}$ and $\overline{SO'}$ will be practically indistinguishable from s and s', so we can rewrite the three sine functions as $\sin \alpha =_p y/s$, $\sin \alpha' =_p y/s'$ and retain $\sin \gamma = y/r$. As angles α, α' and γ are all small in the paraxial case, we can invoke the paraxial approximation again to eliminate the sine functions entirely and write $\alpha =_p y/s$, $\alpha' =_p y/s'$ and $\gamma =_p y/r$. Recalling that in the Cartesian system, $\alpha = \theta$, $\alpha' = -\theta'$, $s = -l$ and $s' = l'$, we can rewrite these as $\theta =_p -y/l$, $-\theta' =_p y/l'$ and $\gamma =_p y/r$.

We can now substitute these expressions for the angles into our rearranged paraxial form of Snell's law, obtaining $\dfrac{n'y}{l'} =_p \dfrac{ny}{l} + \dfrac{(n'-n)y}{r}$.

Interestingly, the ray intercept height y appears in all three terms, so we can divide this out, showing that within the paraxial approximation, the equation holds *independently* of where the incident ray strikes the surface. Hence, we obtain

$$\frac{n'}{l'} =_p \frac{n}{l} + \frac{n'-n}{r} \tag{2.5}$$

which is known as the fundamental paraxial equation.

The simplicity of the single, approximate Equation 2.5, in contrast to the set of seven exact equations (Equation 2.3a-g), begins to illustrate the value of the paraxial approximation. Examination of the three terms in the fundamental paraxial equation reveals further key insights. The first term, n'/l', is the only term that incorporates the image distance l', and the second term, n/l, is the only term that incorporates the object distance l. The third term, $(n' - n)/r$, comprises all of the information about the surface (its radius of curvature and the two refractive indices that define the boundary) and is independent of the object and image locations. Consequently, these three terms are given special recognition. The first, n'/l', is defined as the image vergence L', and the second, n/l, is defined as the object vergence L, where the vergence terminology refers to whether the wavefront is converging (a positive vergence) or diverging (a negative vergence) at the surface. Even more importantly, $(n' - n)/r$ is referred to as the surface power F_{surf}, i.e.,

$$F_{\text{surf}} \equiv \frac{n' - n}{r} \tag{2.6}$$

as this term captures full information (in the paraxial case) about how the surface alters the vergence (convergence or divergence) of the wavefront from the object. With the Cartesian system implemented, the surface power is also robust to changes in the sign of the radius of curvature, so while a convex air-to-glass surface (r positive) will have a positive value of F, a concave air-to-glass surface (r negative) will have a negative F, and these two surface forms will be described as "positive" and "negative" respectively. Similarly, with a glass-to-air surface, such as the back surface of a lens where $n' < n$ and hence $n'-n$ is negative, the surface power F again depends on the sign of r. If the centre of curvature of the second surface is to the left of that surface, as in a biconvex lens, r for that surface will be negative, and along with the negative value for $n'-n$ will make the surface power F positive, as expected for the rear surface of a biconvex lens.

The object vergence, image vergence and surface power all have units of inverse distance, which in SI units is the reciprocal metre, m^{-1}. This unit is known as the dioptre (abbreviation D), and is the common unit for stating optical power in optometry; if you buy a pair of off-the-shelf reading glasses, their strength will almost certainly be marked in dioptres as +1.0 D, +1.5 D, +2.0 D, etc. The human eye, in comparison, has a power of +64 D; it is a high-power optic indeed! The eyepieces we will meet in later chapters have effective powers of several tens to over 100 D.

2.4 REFRACTION BY A LENS WITH SPHERICAL SURFACES

A lens has two optical surfaces, joined by glass. Refraction takes place only at the surfaces, where the refractive index changes; within the lens, the light merely coasts from the first surface to the second. We can therefore calculate the refraction of rays at the second surface using essentially the same set of equations as above, but with one obvious adjustment: we introduced Figure 2.5 for the case of a ray beginning its journey on the optical axis at point O, whereas the ray striking the second surface of a lens began its journey at point S on the first surface, at a height y above the optical axis. However, the ray refracted at the first surface was destined to reach the optical axis at the point O′, at distance l' and angle θ'. If it passes through O′ on its way to the second surface, then as far as the calculation at the second surface is concerned, we can treat the ray as if it did start its journey at O′; in this case, the real image produced by the first surface at O′ becomes the new object point, which we might call O_2, for the second surface. It does not matter at all that the refracted ray actually started its journey at S rather than O′, as the angle that the ray makes upon reaching the second surface (at point S_2 say) and the height it attains at the second surface (y_2, say), is the same for both starting points.

What about a case where the second surface of the lens intercepts the refracted ray *prior* to it reaching the optical axis, as might commonly be the case? Crucially, it does not matter whether the ray from S to O′ actually reaches O′, as the angle of approach to S_2 (θ') is the same in either case, and the distance of the ray from the optical axis, y_2, depends only on that angle of approach and the separation between O′ and V_2.

The importance of this last result cannot be overstated: it is a general characteristic of optics that when considering image formation by sequential surfaces, the image point produced by one surface in the series becomes the object point for the next surface in the set. This is true whether the image of the prior surface is real (formed from converging rays) or virtual (formed from diverging rays), and does not depend on there being sufficient space in the system to form an image within the space prior to the next surface intercepting the light.

After employing these findings and tracing rays through the two surfaces of a lens, there is no reason to stop at that point. We can proceed in the same way to calculate the refraction at a third surface, which may be separated from the second surface by air or by glass of a different type, and to continue in this fashion until an entire optical system has been traversed.

While the set of exact equations (Equation 2.3) can be used recursively to calculate ray paths through the optical surfaces of one or more lenses in a system, there is also value in considering what the fundamental paraxial equation, which we derived in Section 2.3 for a single *surface*, would predict for the behaviour of a *lens*. We do that next.

Paraxial Approximation for a Thin Lens

Recall from Equation 2.5 that under the paraxial approximation, a refracting surface with a radius of curvature r separating media with refractive indices n and n' has object and image distances l and l' related by $\frac{n'}{l'} =_\mathrm{p} \frac{n}{l} + \frac{n'-n}{r}$. When considering a lens, it is helpful to label the two surfaces separately, using subscripts 1 and 2 as follows:

$$\frac{n'_1}{l'_1} =_\mathrm{p} \frac{n_1}{l_1} + \frac{n'_1-n_1}{r_1} \quad \text{and} \quad \frac{n'_2}{l'_2} =_\mathrm{p} \frac{n_2}{l_2} + \frac{n'_2-n_2}{r_2}$$

Utilising the general principle introduced above, that the image produced by the first surface is treated as the object for the second surface, we can infer that the object distance for the second surface, l_2, must be the same as the image distance for the first surface minus the separation d of the two surfaces, i.e., $l_2 = l'_1 - d$. As the refractive index n_2 refers to the same medium as n'_1, i.e., the glass, we can rewrite n_2/l_2 as $n'_1/(l'_1 - d)$. Provided the lens thickness d is small compared to l'_1, we can ignore its role in the denominator and simply write $n_2/l_2 = n'_1/l'_1$; we call such a lens a "thin lens". Substituting this into the fundamental paraxial equation for surface 2 therefore gives us $\frac{n'_2}{l'_2} =_\mathrm{p} \frac{n'_1}{l'_1} + \frac{n'_2-n_2}{r_2}$, and we can replace n'_1/l'_1 by the expression above to obtain

$$\frac{n'_2}{l'_2} =_\mathrm{p} \frac{n_1}{l_1} + \frac{n'_1-n_1}{r_1} + \frac{n'_2-n_2}{r_2} \tag{2.7}$$

For the common case of a thin glass lens in air, we can simplify this further by recognising that $n_1 = n'_2 = 1$, and $n'_1 = n_2 = n_\mathrm{g}$, in which case we can rewrite Equation 2.7 as

$$\frac{1}{l'_2} =_\mathrm{p} \frac{1}{l_1} + \frac{n_\mathrm{g}-1}{r_1} + \frac{1-n_\mathrm{g}}{r_2} \tag{2.8}$$

These last two expressions are especially valuable because l_1 is the distance of the object from the front vertex of the lens, l'_2 is the distance of the image from the back vertex of the lens, and the remaining pair of terms, $(n'_1 - n_1)/r_1 + (n'_2 - n_2)/r_2$, depends only on the characteristics of the lens and evidently corresponds to the power of the lens as a whole. Moreover, this lens-power term is also recognisable as simply the sum of the two surface powers, i.e.,

$$F_{\text{thinlens}} = F_{\text{surf1}} + F_{\text{surf2}} \qquad (2.9)$$

Evidently, when paraxial rays pass through a thin lens, the lens performs simply as the sum of the two surface powers.

An important implication of this simple sum can be noted by recognising that there are many combinations of surface powers that will give the same overall thin-lens power. For example, a +4 D thin lens could be made as an equiconvex lens having two +2 D surfaces, or as a plano-convex lens with surface powers 0 D and +4 D, or conversely +4 D and 0 D, or as a meniscus lens with surface powers +6 D and −2 D. Many other pairings are possible. While all such combinations have the same total power, their performance for non-paraxial rays will differ, so changing the surface power pairings, while keeping the overall lens power fixed, provides the lens designer with one important means of optimising the imaging performance of a lens within the constraint of its required overall power. This approach, based as it is on varying the two surface powers and hence radii of curvature in tandem, is referred to as "bending" a lens. We shall explore lens bending further in later chapters.

2.5 POWERS AND FOCAL LENGTHS

So far, we have defined surface powers (Equation 2.6) and thin-lens powers (Equations 2.7–2.9) without discussing how these relate to the focal lengths of the optics. We address this topic here.

Rays approaching a positive optical surface or lens from a very distant object point (essentially at infinity) on the optical axis, travelling from left to right as per the usual convention, will reach the surface or lens travelling parallel to the optical axis, and will then be refracted towards the optical axis. All rays that meet the paraxial condition, i.e., that the angle of incidence is very small, will pass through the *same* image point at distance l' on the optical axis. (Recall from the derivation of the fundamental paraxial equation that the image distance l' does not depend on the height y at which the rays strike the surface.) This image point is called the focal point (strictly the *second* focal point) of the surface or lens. Rays that do not satisfy the paraxial condition will not pass through the same point, and for such rays, the notion of a focal point is slightly more tenuous; we shall return to that case in Chapter 3.

We can use the fundamental paraxial equation for a surface and for a thin lens so see how their focal lengths relate to their focal power.

Relationship Between Focal Power and Focal Length

Starting with the fundamental paraxial equation for a surface, $n'/l' =_p n/l + F_{surf}$ (Equation 2.5), if we set the object distance l to negative infinity, then the term n/l becomes zero, and we find that the image location l' must satisfy the relation $n'/l' =_p F_{surf}$, or equivalently $l' = n'/F_{surf}$. This image location for rays approaching from infinity is exactly what we mean when we use the phrase "(second) focal length", so the symbol l' in this case is replaced with a special symbol used just for this case, f', meaning the second focal length of the surface. Hence, for a surface, the second focal length

$$f' = n'/F_{surf} \tag{2.10}$$

Starting instead with the fundamental paraxial equation for a thin *lens*, $n'_2/l'_2 =_p n_1/l_1 + F_{thinlens}$ (Equations 2.7, 2.6 and 2.9), we can again set the object distance l_1 to negative infinity, and find $n'_2/l'_2 =_p F_{thinlens}$ in which case $l'_2 = n'_2/F_{thinlens}$. Recognising again that this is a very special case, we relabel this particular value of l'_2 as the (second) focal length f' of the lens, but additionally for a lens in air, $n'_2 = 1.00$, so we can conclude that

$$f' = 1/F_{thinlens} \tag{2.11}$$

As an example, a thin lens of power 2 D has a second focal length of 0.5 m.

The text above has emphasised the importance of recognising that the focal lengths discussed so far are *second* focal lengths, but what of the *first* focal length, which so far has not been mentioned? We set up the derivation of the second focal length by considering a distant object giving rise to rays approaching the surface or lens from the left, parallel to the optical axis. We can also consider a different pencil of rays leaving the surface or lens towards the right, parallel to the optical axis, and ask where the source point (object point) for such a set of rays must lie. The fundamental paraxial equations for a surface and for a thin lens again answer this question. Rays exiting the surface or lens parallel to the optical axis correspond to an image formed at infinity. Starting with the fundamental paraxial equation for a surface, $n'/l' =_p n/l + F_{surf}$ (Equation 2.5), if we set the image distance l' to infinity, then the term n'/l' becomes zero, and we find that the object location l must satisfy the relation $n/l + F_{surf} =_p 0$, or equivalently

$l =_p - n/F_{surf}$. The source point (object point) for rays exiting the surface parallel to the optical axis is exactly what we mean when we use the phrase "first focal length", so the symbol l in this case is replaced with a special symbol used just for this case, f, meaning the first focal length. Hence, for a surface, the first focal length

$$f = -n / F_{surf} \qquad (2.12)$$

Starting instead with the fundamental paraxial equation for a thin lens, $n'_2/l'_2 =_p n_1/l_1 + F_{thinlens}$ (Equation 2.7, 2.6 and 2.9), we can again set the image distance l'_2 to infinity, and find $n_1/l_1 + F_{thinlens} =_p 0$ in which case $l_1 = -n'_2/F_{thinlens}$. Recognising again that this is a very special case, we relabel this particular value of l_1 as the first focal length f, but again note that for a lens in air, $n'_2 = 1.00$, so we can conclude that

$$f = -1 / F_{thinlens} \qquad (2.13)$$

Note that for the thin lens, $f = -f'$ so the two focal points of the thin lens are equidistant from the lens and on opposite sides, but for a surface they are not equidistant.

These relationships between focal lengths and focal powers apply even if the surfaces or lenses are negative powered. In such cases, the focal powers will be negative, so the second focal length f' will also be negative, i.e. to the left, and the first focal length f will be positive. Positive and negative surfaces and lenses, their first and second focal points F and F', and their first and second focal lengths f and f', are illustrated in Figure 2.6.

The derivation of the fundamental paraxial equation for a thin lens, set out above, made the approximation that the lens thickness d was negligible compared to the distance l'_1 to the image formed by the first surface. This greatly simplified the algebra, but the lens thickness could have been retained in the working, which would have yielded the focal power and focal lengths for a thick lens. That derivation would have taken longer than is useful for our purposes, so the results will simply be stated below; the full derivation may be found elsewhere.[26]

The focal points of a thick lens are only symmetric about the centre plane of the lens if the lens itself is symmetric, i.e., if the first and second surfaces have the same power, and if both surfaces are in contact with the same external medium, as is the case for a symmetric thick lens in air, say. More generally, the first and second focal lengths are different. Moreover,

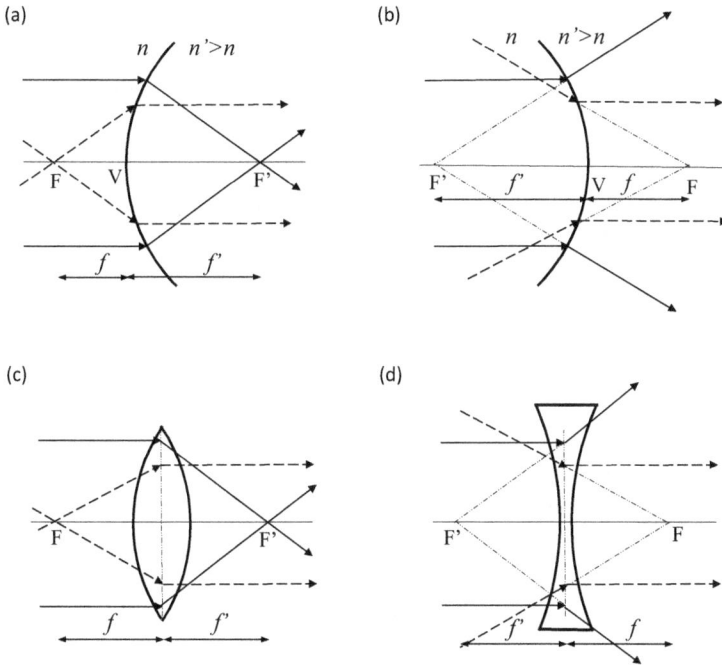

FIGURE 2.6 First and second focal points and focal lengths of positive and negative surfaces and thin lenses. Schematic diagrams (not to scale) showing the locations of the first and second focal points F and F′, and the first and second focal lengths f and $f′$, of (a) a positive air-to-glass surface, (b) a negative air-to-glass surface, (c) a positive thin lens, and (d) a negative thin lens. The focal lengths are not to scale; typically for an air-to-glass surface with $n_{glass} \approx 1.5$, $f′ \approx 3r$ and $f \approx -2r$, and for a symmetric thin lens, $f′ \approx 2r$ and $f = -f′$. Rays associated with the first focal point are shown as dashed lines, while rays associated with the second focal point are shown unbroken. For the negative surface (b) and negative thin lens (d), additional construction lines (dot-dash) are provided since the rays do not physically pass through the focal points. For the thin lens, the rays are shown to be refracted at the centre plane of the lens, since a thin lens by definition has negligible thickness.

whereas a single surface and a thin lens have one obvious reference point along the optical axis, the situation is more complex for a thick lens due to the separation of its surfaces, and as a consequence, a thick lens has three focal powers and four focal lengths! For a thick lens in air, these are:

- The effective focal power and corresponding focal lengths given by

$$F_E = F_1 + F_2 - (d/n)F_1F_2 \quad \text{and} \quad f′_E = 1/F_E = -f_E \qquad (2.14)$$

where n is the refractive index of the lens material. Note that if the lens thickness $d = 0$, then this reduces to the expression we have already met for the thin lens, as you would expect. The effective focal length $f'_E = 1/F_E$, with the prime (′) in the notation, is a second effective focal length, while the corresponding first effective focal length $f_E = -f'_E$. The first and second effective focal lengths give the location of the focal points relative to two reference planes called the first and second principal planes, P and P′; the distance of F from P is f_E, and the distance of F′ from P′ is f'_E. The significance of the principal planes is that, in the paraxial approximation, rays incident from the left and travelling parallel to the optical axis will be refracted through the second focal point F′ as if they experienced just one refraction at P′. Similarly, paraxial rays exiting the lens towards the right, parallel to the optical axis, will have approached the lens through the first focal point F at an angle consistent with undergoing a single refraction in the plane P. Of course, all rays undergo two refractions, at the first and second air-glass and glass-air surfaces of the lens, but the principal plane analysis allows us to imagine and calculate ray paths for paraxial rays as if only one refraction took place, which can simplify matters. The planes P and P′ are separated by a small distance along the optical axis called the hiatus (see Section 2.6), which is usually comparable to the thickness of the lens, and the ray paths can effectively be ignored between the two planes; they are rendered horizontal in diagrams if drawn at all. Moreover, and this is perhaps the greatest benefit of the concept, the principal plane concept also applies to systems of many lenses, so it is common to quote the effective focal lengths of telescopes, camera lenses and eyepieces so their behaviour is summarised by a single focal length, without the user needing to know the details of the many lens elements that may be embedded within the system.

- In addition to the effective focal length f'_E (and f_E), a thick lens is also characterised by the distance of the second focal point F′ from the back vertex of the lens (V_2), and by the distance of the first focal point from the front vertex V_1. These focal lengths we label as f'_V and f_V, respectively. The second vertex focal length f'_V can be calculated from the second vertex power F'_V via

$$F'_V = \frac{F_E}{1 - \dfrac{d}{n}F_1} \text{ and } f'_V = 1/F'_V \tag{2.15}$$

while the first vertex focal length f_V can be calculated from the first vertex power F_V via

$$F_V = \frac{F_E}{1 - \dfrac{d}{n}F_2} \text{ and } f_V = -1/F_V \qquad (2.16)$$

Temporarily setting the lens thickness d to zero would imply $F_V = F_V' = F_E$, and the two focal points would again be symmetric about the lens centre with $f_V = -f_V'$ as expected for a thin lens, but in general, they will not be symmetric in an asymmetric thick lens. The second vertex focal length is sometimes referred to as the back focal distance, and it can be an important consideration for a lens or optical system to ensure there is sufficient space between the lens and focal point if other optical elements or mechanical items (such as a shutter) need to be included in that part of the optical train.

It is useful to note that the thick-lens equations, including the effective focal power (Equation 2.14) can be re-purposed for a pair of thin lenses in air separated by a distance d if we recognise n as the refractive index of the medium between the two refractive elements (lenses), which in this case is air having $n = 1$. Thus, for two separated thin lenses in air, we can write

$$F_E = F_1 + F_2 - dF_1F_2 \qquad (2.17)$$

and $f_E' = 1/F_E$, etc., as before.

2.6 NODAL POINTS, APERTURES AND PUPILS

Rays arriving at the centre of a thin lens pass through the lens undeviated (Figure 2.7a), even if they arrive inclined to the optical axis, as is the case for rays originating from objects off the optical axis. This fortuitous situation arises because the two surfaces of a lens are parallel at the two vertices V_1 and V_2, and in a thin lens, having zero thickness, there is no sideways deviation of the light between V_1 and V_2. Such rays are of course principal rays if the lens also serves as the aperture stop (see Section 1.1). The ability to trace principal rays from a range of object points passing undeviated through a thin lens greatly simplifies the task of tracing their propagation.

For thick lenses and lens systems, the identification of undeviated rays is more complex. Fortunately, for paraxial rays, we can still identify a point

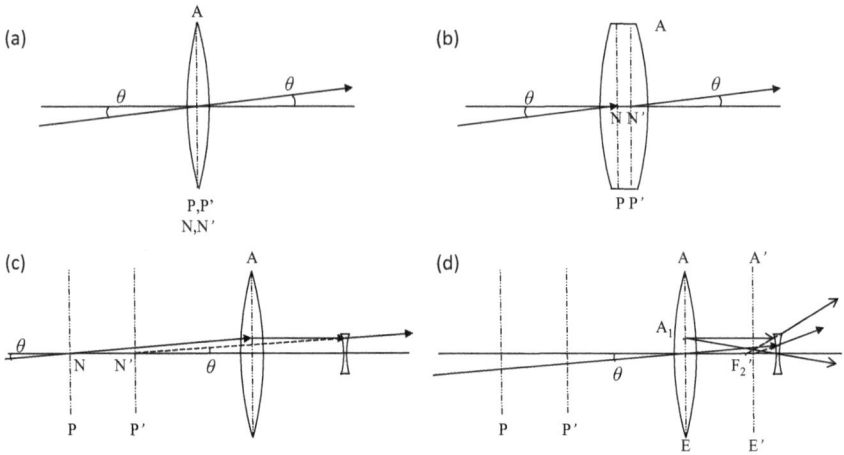

FIGURE 2.7 Nodal points of lenses and lens systems (schematic diagram, not to scale). (a) positive thin lens: the principal planes P and P′ and the nodal points N and N′ all lie in the central plane of a lens of negligible thickness, and a ray incident upon the centre of the lens at an any angle θ to the optical axis is undeviated. The principal planes coincide with the aperture stop A that is defined by the edge of the lens. (b) biconvex thick lens: the principal planes P and P′ lie within the lens, separated by approximately $1/3^{rd}$ of the lens thickness, and the nodal points N and N′ are at the intersections of the principal planes with the optical axis. A ray directed towards the first nodal point N at any paraxial angle θ to the optical axis will be refracted at both the first and second surfaces of the lens, but ultimately will exit the lens as if it had come from the second nodal point N′ at the same angle θ. The aperture stop A is defined by the edge of the lens. (c) Thin-lens equivalent of a Cassegrain telescope: the combination of positive and negative optics moves the principal planes considerably forward of the space occupied by the optics themselves, and the principal planes no longer coincide with the aperture stop defined by the primary optic. Nevertheless, nodal points N and N′ still exist where the principal planes intersect the optical axis; for a distant object lying off-axis at any paraxial angle θ to the optical axis, the ray that happens to pass through the first nodal point N will ultimately exit the telescope as if it had exited from the second nodal point N′ and will do so at the same angle θ. This ray will not be a principal ray, as it will not have passed through the centre of the aperture stop A, but can usefully be referred to as a nodal ray. (d) As for (c), but showing the ray construction to determine where the image A′ of the aperture stop A is located as a result of refraction by the negative secondary optic, and showing the path of a principal ray which enters the system in the centre of the aperture stop A and consequently exits the system as if it has emerged from the centre of A′. In this instance, A is also the entrance pupil E of the system, and the image A′ of the entrance aperture A (and entrance pupil E) formed by later optics in the system is by definition the exit pupil E′.

on the optical axis through which incoming rays can be directed and then be found to emerge as if from a second point on the optical axis, *at the same angle*, and thus undeviated in angle albeit displaced laterally. The point through which incident paraxial rays must be directed in order to emerge at the same angle to the optical axis is called the first nodal point N, and the point on the optical axis from which they appear to emerge when they exit the lens or system is called the second nodal point N'. For a lens or lens system in air, the first and second nodal points correspond to the first and second principal points, which are where the first and second principal planes intersect the optical axis (Figure 2.7b and c).

Although the principal planes and nodal points were introduced above in connection with a thick lens, they still exist for a thin lens but are simply co-located in the centre of a lens having zero thickness. For a thick lens, the two principal planes separate, and so the nodal points separate with them. Their separation, known at the "hiatus" of the lens, is typically around $1/3^{rd}$ the thickness of a lens.[27]

In contrast to the case for a thin lens, the principal planes of a thick lens need not be located within the lens. In particular, the principal planes of meniscus lenses, which have one positive and one negative surface, often lie outside the lens. One obvious implication is that the aperture stop of the lens or system may then be separated from the principal plane, although for a thick biconvex or biconcave lens, the two will at least be close unless a separate aperture stop is instituted. Two examples of separated aperture stops are found in the classical Schmidt camera and early photographic "landscape" lenses. The aperture stop of the classical Schmidt camera is placed at the centre of curvature of a primary spherical mirror; sub-portions of the oversized spherical mirror are illuminated differently by sources located in different directions, but as the principal ray of every source necessarily passes through the centre of curvature, the path of every principal ray constitutes an axis of symmetry of the spherical mirror. Consequently, similar image quality is obtained across the field of view and, as will make more sense by the end of Chapter 3, there are no off-axis aberrations.[28,29] Simple landscape lenses for early cameras were devised based on a single meniscus lens and a displaced stop which, incidentally, was on the opposite side of the lens to the principal plane.[30]

In a refracting telescope or Newtonian reflecting telescope, the aperture stop is essentially coincident with the major optic, and the principal planes and nodal points will also coincide with it. The principal rays of objects off the optical axis will therefore pass more or less through the nodal points and thus travel on to the focal plane at the same angle. In telescopes of Cassegrain and similar configurations, however, a convex secondary mirror effectively pushes the second principal plane beyond the primary mirror and telescope tube assembly, and thus separates it from the aperture stop of the telescope (see Figure 2.7c for a schematic thin-lens simulation of a Cassegrain configuration[31]). Nevertheless, because the field of view of an astronomical telescope is invariably small (typically ~1°) and the objects are at a great distance, within the pencil of rays coming from an object located an angle θ off the optical axis could be one ray directed through the first nodal point N, which will reach the focal plane as if it had come from the second nodal point N′, where the second principal plane P′ cuts the optical axis, at the same angle θ. (Recall that the second principal plane lies one effective focal length upstream of the focal plane.) This nodal ray is not a principal ray, as it has not passed through the centre of the aperture stop, but the principal ray will nevertheless converge with the nodal ray in the focal plane, and therefore the nodal ray informs us where in the focal plane the principal ray is to be found.

The path taken by the principal ray can be found by considering the role of the second, negative lens, which mimics the negative secondary mirror of a Cassegrain telescope in our thin-lens-equivalent diagram (Figure 2.7c). This secondary optic produces a virtual image of the primary optic, which appears slightly upstream of the secondary, reduced substantially in size. This can be seen in Figure 2.7d, which shows the same optics as panel (c) but with the addition of a point A_1 in the aperture stop and the second focal point of the secondary optic, F_2'. (As the secondary optic is negative, the second focal length is also negative, and hence F_2' lies upstream of the lens.) To reveal the location of the image A′ of the aperture stop A formed by the secondary optic, we trace two rays from A_1: the first is drawn through the centre of the secondary thin lens where it proceeds undeviated, and the second is drawn parallel to the optical axis so that when it is refracted by the negative second lens, it will appear to have come from its second focal point, F_2'. Tracing this refracted ray back to its apparent intersection with the first, undeviated ray indicates where the image of A_1 will be formed, and more generally reveals the location of the image A′ of the aperture stop A. As the principal ray enters the telescope in the centre of the aperture stop A, it will exit the system as if emerging from the centre of the image of the aperture stop A′, as drawn in Figure 2.7d.

The importance of these points is recognised by introducing a distinct terminology. The entrance aperture A in this instance is also known as the entrance pupil E of the telescope, and the image A′ of the entrance aperture A (and E) formed by subsequent optics is called the exit pupil E′. In a refracting telescope or Newtonian telescope, the exit pupil of the telescope optics (i.e. excluding the eyepiece) is located at the primary optic, but in Cassegrain and similar systems where a negative secondary mirror further modifies the ray path, as simulated in Figure 2.7c and d where a negative lens is shown fulfilling this role, the exit pupil E′ is located close to and just "behind" the negative secondary optic (i.e. "behind" from the perspective of an observer looking up the optical path from the eyepiece end without an eyepiece inserted). The location of the exit pupil can be calculated exactly using the fundamental paraxial equation if the radius of curvature (or focal length or power) of the secondary mirror is known, treating the entrance aperture A of the telescope as the object.

The image of the entrance aperture seen at E′ is smaller than the original aperture A, as it has been imaged by a negative optic. Nevertheless, the smaller, closer exit pupil E′ must still deliver pencils to the telescope focal plane consistent with the system f/ratio, given as usual by the effective focal length of the telescope f'_E divided by the diameter D of the aperture A: f/ratio $= f'_E/D$. One difference, however, in a system having a negative secondary optic is that although principal rays enter the entrance aperture at some small angle θ to the optical axis, they will exit the exit pupil at a greater angle θ', where the magnification factor θ'/θ is given by $\dfrac{f/\text{ratio}_{\text{telescope}}}{f/\text{ratio}_{\text{primaryoptic}}}$, which is often around 10/2 or similar,[32] i.e. θ' ~ 5θ. When it comes to examining eyepieces, this means the angles at which principal rays enter the eyepiece depend on the distance to the exit pupil of the telescope (without eyepiece) to which it is to be attached, so we have to make sensible assumptions about this. For computational convenience, in the subsequent chapters, we simulate f/10 pencils assuming a 100 mm diameter exit pupil located 1000 mm upstream of the telescope focal plane, and we simulate f/6 pencils using a 167 mm diameter exit pupil at the same distance.

Using all this information, we can use trigonometry to calculate the trajectory of both the nodal ray emerging from N′ at the original angle θ and the principal ray emerging from the centre of E′ at its angle θ', to trace both rays to their intersection in the focal plane of the telescope. The calculation of the refracted angle of the principal ray depends on the powers and separation of the primary and secondary optics, i.e. F_1, F_2 and d, but is simply an application of the separated thin-lens-pair equation (Equation 2.17) already introduced above.

2.7 LENS ANATOMY

When describing eyepieces, we will talk about systems comprising multiple lenses, some of which are in contact and some of which are separated. The terms "element", "component" and "group" constitute a hierarchy of manufacture and organisation within an optical "system".

An "element" or "lens element" refers to any single piece of glass polished as a lens having two refracting surfaces. Occasionally, two or three elements may be placed in contact and bonded with transparent optical cement to form a composite unit, commonly called a "component", which may be picked up as one piece. A component comprising two elements is usually called a doublet, and a component comprising three elements is a triplet. In a slight variation, the elements of a doublet or triplet may be separated by a narrow airspace rather than in full contact, but may still be regarded as a doublet or triplet because of their close physical and optical relationship.

A "group" consists of one or more elements and/or components that are located more or less together, but not necessarily bonded, and which appear spatially and/or functionally distinct from some other group (which likewise consists of one or more elements and/or components).

There is no ambiguity over the use of the terms element and component, but what exactly constitutes a group is much more subjective, and for a system of adjacent lenses, it is often unclear where one group ends and another begins. The "group" terminology should therefore be treated as a phrase of convenience, and it is not usually worth debating the merits or otherwise of a particular "grouping" of lenses. This issue will frequently appear when we discuss multi-lens systems and encounter the imprecise "group" terminology describing the relationships of lenses, for example, where a single component in one design may be split into two components in another, where they may or may not be described as belonging to the same group.

We have now acquired an understanding of how and why light is refracted at an optical surface, how rays can be traced exactly through one and more optical surfaces, and how the paraxial approximation and the thin-lens approximation provide us with more intuitive insights and help us form expectations about the gross performance of a lens or lens system of known characteristics. To bring these concepts to life, we can now proceed to model the performance of a single lens as an eyepiece and better understand the issues introduced above by applying them. In the next chapter, we will therefore begin to quantify the performance of a single-lens eyepiece and identify its limitations, as a prelude to exploring improved designs in later chapters.

NOTES

1 M.H. Freeman and C.C. Hull, *Optics*, 11th edition, Butterworth Heinemann 2003, Chapter 3.
2 M.H. Freeman and C.C. Hull, *Optics*, 11th edition, Butterworth Heinemann 2003, Chapter 10.
3 Schott datasheet N-BK7, 2023 https://media.schott.com/api/public/content/41e799d0bf874807a0bb8e702fbb75b5?v=54856406 (accessed 18/01/2025).
4 Schott datasheet N-BAK4, 2014 https://media.schott.com/api/public/content/06c6b5a65e0c417d98008ec5f7b0daa5?v=b6999a17 (accessed 18/01/2025).
5 Schott datasheet N-F2, 2014 https://media.schott.com/api/public/content/061f3156c83a44ed9220770b0f65a869?v=d69b35e0 (accessed 18/01/2025).
6 Schott datasheet N-SF11, 2020 https://media.schott.com/api/public/content/78e83df5ca2c4da4ad4490a52c80a146?v=1a468147 (accessed 18/01/2025).
7 F.A. Jenkins and H.E. White, *Fundamentals of Optics*, 4th edition, McGraw-Hill 1976, Chapter 1.
8 F.A. Jenkins and H.E. White, *Fundamentals of Optics*, 4th edition, McGraw-Hill 1976, Chapter 1.
9 M.H. Freeman and C.C. Hull, *Optics*, 11th edition, Butterworth Heinemann 2003, Chapter 3.
10 M.H. Freeman and C.C. Hull, *Optics*, 11th edition, Butterworth Heinemann 2003, Chapter 11.
11 M.J. Kidger, *Fundamental Optical Design*, SPIE, 2000, Chapter 2.
12 M.H. Freeman and C.C. Hull, *Optics*, 11th edition, Butterworth Heinemann 2003, Chapter 10.
13 F.A. Jenkins and H.E. White, *Fundamentals of Optics*, 4th edition, McGraw-Hill 1976, Chapter 23.
14 Thorlabs' optical glass datasheets https://www.thorlabs.com/newgroup-page9.cfm?objectgroup_id=6973&tabname=N-BK7 (accessed 18/01/2025).
15 M.H. Freeman and C.C. Hull, *Optics*, 11th edition, Butterworth Heinemann 2003, Chapter 3.
16 R. Kingslake and R.B. Johnson, *Lens Design Fundamentals*, 2010, Academic Press, Chapter 2.
17 M.J. Kidger, *Fundamental Optical Design*, SPIE, 2000, Chapter 2.
18 F.A. Jenkins and H.E. White, *Fundamentals of Optics*, 4th edition, McGraw-Hill 1976, Chapter 3.
19 E. Hecht, *Optics*, 5th edition, Pearsons 2017, Chapter 5.
20 M.H. Freeman and C.C. Hull, *Optics*, 11th edition, Butterworth Heinemann 2003, Chapter 3.

21 R. Kingslake and R.B. Johnson, *Lens Design Fundamentals*, 2010, Academic Press, Chapter 2.

22 R. Kingslake and R.B. Johnson, *Lens Design Fundamentals*, 2010, Academic Press, Chapter 2.

23 Kingslake and Johnson (see endnote 22) likewise adopt the Cartesian convention, but they construct ray-tracing equations based on the impact parameter Q of a ray approaching the surface. They also consider more advanced cases such as aspheric surfaces which we will not consider in such detail.

24 A radian is an angular measure defined such that there are 2π radians (i.e. approximately 6.3 radians) in 360°. 1 radian is therefore approximately 57.3°. This definition has the convenience that on a circle of radius r, an arc length s subtends an angle θ in radians at the centre, and the three quantities are related as $s = r\theta$.

25 The mathematical symbol "$n!$" is read as "n factorial", and signifies the product of all positive integers up to and including n. For example, $4! = 1 \times 2 \times 3 \times 4 = 24$.

26 M.H. Freeman and C.C. Hull, *Optics*, 11th edition, Butterworth Heinemann 2003, Chapter 5.

27 R. Kingslake and R.B. Johnson, *Lens Design Fundamentals*, 2010, Academic Press, Chapter 3.

28 M.J. Kidger, *Fundamental Optical Design*, 2000, SPIE, Chapter 13.

29 E. Hecht, *Optics*, 5th edition, 2017, Pearsons, Chapter 5.

30 R. Kingslake and R.B. Johnson, *Lens Design Fundamentals*, 2010, Academic Press, Chapter 12.

31 See also the analysis of a two-lens telephoto lens in M.H. Freeman and C.C. Hull, *Optics*, 11th edition, Butterworth Heinemann 2003, Chapter 6.

32 D.J. Schroeder, *Astronomical Optics*, 2nd edition, Academic Press, 2000, Chapter 2.

CHAPTER 3

A Single-Lens Eyepiece and Aberrations

3.1 DESIGN CONSTRAINTS

In this chapter, we embark on a design effort that, initially at least, will be of very limited success. We assume we have a telescope of unspecified optical design that has an entrance aperture diameter D = 300 mm and an effective focal length f'_{tel} = 3000 mm, and thus operates at $f/10$. We assume the telescope (excluding the eyepiece) has an exit pupil 1000 mm upstream of the focal plane, with a diameter of 100 mm. Somewhat optimistically, we further assume that it forms a perfect, flat, real image in its focal plane. We can infer that:

- The image scale in the real image will be 57.3°/3000 mm = 0.019°/mm, or 1.15 arcmin mm^{-1}.

- The resolution limit of the telescope, i.e., the Airy disk radius given in radians by $1.22\lambda/D$, for a typical visual wavelength of 550 nm, is 2.2×10^{-6} radians or 0.46 arcsec.

We will adopt the following desired design criteria for a simple eyepiece, which is to be used to magnify the image from this telescope:

- Glass: The eyepiece will comprise a single glass lens of a commonly available optical glass. For this, we will adopt N-BK7, which is one of the most commonly used optical glasses; it has a refractive index of

DOI: 10.1201/9781003670506-3

1.5168 at a common visual reference wavelength of 587.6 nm (which corresponds to the "D3" or "d" spectral line[1,2] of neutral helium). For comparison, $n_{air} = 1.0003$.

- Infinity adjustment: The eyepiece is to be used in "infinity adjustment", which is to say it will produce a virtual image at infinity. This is characterised by the rays within any given pencil travelling parallel to one another when they exit the eyepiece. Focussing of the human eye is called accommodation and is achieved by adjusting the tension in the ciliary muscles, which pull the forward (anterior) surface of the flexible lens of the human eye into a more strongly curved shape to focus on nearer objects. Using a telescope in infinity adjustment is advantageous because the ciliary muscles that focus the eye are more or less relaxed when viewing an object or image at infinity (but see endnotes[3,4]) and are more comfortable in this configuration than if they were constantly tensioned to hold the eye in focus at an intermediate distance. Of course, the plan to work in infinity adjustment breaks down slightly for observers having myopia (short-sightedness), who either need to wear spectacles to be able to focus on an image at infinity, or else need to adjust the focus of the telescope to generate slightly diverging pencils that their over-powered eyes can bring to focus.

- Angular magnification: The angular magnification of a telescope used visually refers to the ratio of the angle that the principal ray of an off-axis object makes relative to the optical axis when it exits the eyepiece, θ', compared to when it entered the telescope, θ, which in the paraxial case (thus ignoring distortion; Section 3.7) is given by the ratio of focal lengths:

$$m_{ang} \equiv \theta' / \theta =_p f'_{tel} / f'_{eye} \tag{3.1}$$

For our design, we desire an angular magnification $m_{ang} = 120$, which requires an eyepiece focal length $f'_{eye} =_p f'_{tel} / m_{ang} = 3000 \text{ mm}/120 = 25 \text{ mm}$.

- Eye relief: The eyepiece, being a positive lens or positive system of lenses, produces an image of the entrance aperture of the telescope behind the eyepiece. (Equivalently, it forms an image of the exit pupil of the telescope's main optics, which is in turn an image of the entrance aperture.) The image of the telescope's aperture (and entrance pupil and main-optics' exit pupil) formed by the eyepiece is consequently the exit pupil of the entire telescope+eyepiece system, and its significance is that every ray of light entering the telescope emerges via the exit pupil. In particular, rays passing through the centre of the aperture, which we call principal rays, will appear to exit from the centre of the exit pupil (at least in the absence of pupil aberrations, which we consider later in the book). In our imaginary telescope, the exit pupil of the main optics is 1000 mm upstream of the eyepiece. This is far compared to the focal length of the eyepiece (25 mm), so the system exit pupil is formed approximately one eyepiece focal length downstream of the single-lens eyepiece. Importantly for a visual observer, the entrance pupil of the human eye (which we typically just call the pupil) should be co-located with the exit pupil of the telescope+eyepiece, so that every ray entering the telescope aperture and then exiting at the telescope+eyepiece exit pupil will be conveyed into the eye. The distance between the back surface of the eyepiece lens and the system exit pupil is called the eye relief. If the eye relief is too small, typically less than around 10 mm, it becomes very challenging for the observer to position their eye close enough to the eyepiece to make the eye's pupil and the telescope+eyepiece exit pupil coincide. We shall have to be careful not to design an eyepiece having too small an eye relief.

- Field of view: Ideally, our telescope would be capable of observing the full diameter of the Moon at closest approach to the Earth, which is 34 arcmin across. At the image scale calculated above, i.e., 1.15 arcmin mm^{-1}, this implies a field of view 29.8 mm across, which just exceeds the ~28 mm inside diameter of the barrel of many 1¼ inch standard eyepieces. We will therefore target a 28 mm diameter field of view, i.e., 32 arcmin. A 28 mm diameter field observed with a 25 mm focal length eyepiece implies an apparent field of view of 29° either side of the optical axis (ignoring any distortion; see Section 3.7), so we should aim to achieve a 60° apparent field of view.

It's all right to wish!

3.2 THIN LENS AND THICK LENS DESIGNS

A thin lens of focal length $f' = 25$ mm requires a focal power $F = 1/0.025$ m $= 40$ D. There are many thin lens shapes that could deliver this, so we will consider two obvious possibilities:

- a symmetric thin lens having two 20 D surfaces, since $F_{thinlens} = F_1 + F_2$
- a plano-convex lens with one flat surface ($F_1 = 0$ D) and the other convex ($F_2 = 40$ D)

A single-lens eyepiece of 25 mm focal length would have to be located 25 mm beyond the telescope focal plane to achieve infinity adjustment. Over that distance, the $f/10$ ray pencils for each object point will expand to a diameter of 25 mm/10 = 2.5 mm. Also, points at the edge of our 32 arcmin field of view, i.e., 14 mm from the optical axis, have principal rays coming from the centre of the exit pupil of the main optics, which we have stipulated (Section 3.1) is 1000 mm away. These principal rays are therefore diverging from the optical axis and will travel a further 0.35 mm away from the optical axis when they reach the eyepiece lens. The clear aperture of the eyepiece lens needs to be large enough to intercept the diverging principal rays from the edge of the field of view and to intercept the expanding pencil of rays for each object point, so our lens needs a clear radius of 14 + 0.35 + (2.5/2) mm = 15.6 mm. An additional 1 mm is required to mount the lens, and another 1 mm beyond that for trimming during manufacture, so we should assume an initial physical radius of 17.6 mm.[5]

In a symmetric N-BK7 lens having surface powers of 20 D, the radius of curvature must satisfy $F_{surf} = (n' - n)/r$, i.e., $r = 0.5165/20 = 0.0258$ m = 25.8 mm. The central "bulge" or sag ζ of a spherical surface (Figure 3.1) of physical radius y and radius of curvature r is given by application of Pythagoras' theorem as

$$\zeta = r - \sqrt{r^2 - y^2} \qquad (3.2)$$

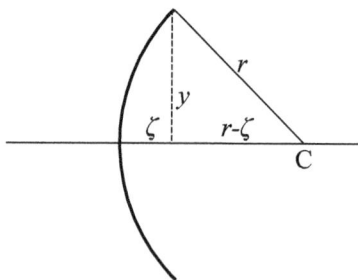

FIGURE 3.1 Geometry of a spherical surface. Cross-section of a spherical surface having a radius of curvature r measured from the centre of curvature C, and having a physical radius y, protrudes a distance ζ from a plane passing through the edge of the surface. The distance of protrusion, called the sag, is obtained with the aid of Pythagoras' theorem as in Equation 3.2.

For our lens surface with r = 25.8 mm and physical radius y = 17.6 mm, we infer a sag of 6.9 mm on each surface. Adopting a minimum edge thickness of 1 mm,[6] we therefore infer a minimum required thickness of 15 mm for our eyepiece lens.

Given this minimum thickness d = 15 mm, we can now use the thick lens equation to see what the non-zero thickness implies for the surface powers of the lens. For a symmetric thick lens, where F_1 = F_2, we infer that stronger surface powers ~22.5 D will be required, rather than 20 D as inferred from the thin lens equation, to deliver F = 40 D. Stronger surfaces imply greater sags and a still thicker lens; further cycles of calculations imply F_1 = 23.27 D, r_1 = 22.19 mm, and a thickness of 18.3 mm for a 25 mm focal length, symmetric biconvex lens. The front vertex focal length of this lens, which will become the distance between the real image formed by the main telescope optics and the front vertex of the lens, is given by Equation 2.16, from which we obtain −18.0 mm. The second vertex focal length for this symmetric lens will therefore be +18.0 mm, and the eye relief will be close to this value too, ~18.6 mm.

In contrast, a plano-convex lens option is readily excluded. The convex surface would require a power of 40 D, which for a glass of refractive index 1.5 implies a radius of curvature r = 12.5 mm, which cannot deliver the required lens radius of 17.6 mm. We could increase the radius of curvature by choosing a glass with a higher refractive index, but we might as well abandon this lens already because of the strong surface curvatures implied.

At this point, it would be useful to define mathematically the *curvature* $1/r$, which is the reciprocal of r, the *radius of curvature*. A surface with a small radius of curvature is obviously very strongly curved and thus has a high curvature, while a surface with a long radius of curvature may appear almost flat, and hence its curvature is small. It is important to distinguish carefully between the two concepts in both words and equations. (Many people shorten "radius of curvature" to "radius", which I do not encourage since it can be confused with the physical radius of a circular lens.)

We will therefore investigate just one option here: a symmetric biconvex lens with surface radii of curvature r_1 = +22.19 mm and r_2 = −22.19 mm, a central thickness of 18.3 mm, and a clear aperture of 31.2 mm, made of N-BK7 glass of refractive index 1.5168.

3.3 RAY TRACING THE LENS

We begin by calculating a ray trace from an on-axis object point in the real image formed by the main optics of the telescope, using the set of exact equations in Equations 2.3a-g. In an $f/10$ pencil, the rays make angles up to 2.9° from the optical axis, so we will trace indicative rays at 0.1°, 1.0°, 2.0° and 3.0°. The first of these is to simulate a paraxial ray, while the others indicate more inclined rays. Ideally, we would find the paraxial ray exiting the lens parallel to the optical axis, since our intention is to operate in infinity adjustment. In practice, a little tweaking of the input parameters is required to achieve this; one option, which is quite practical in systems involving multiple surfaces, is to refine the radius of curvature of the last surface to obtain the desired ray angle,[7] but in our case where we have a single lens we can readily shift it slightly along the optical axis, from 18.0 mm as estimated above to 17.98 mm, as one would do in practice when focussing a telescope to focus the paraxial rays. The results of the calculation are set out one ray at a time and one surface at a time in Figure 3.2a. While the emerging paraxial ray angle is close to zero as required (<0.1 arcsec), the rays in the pencil at 1°, 2° and 3° to the optical axis are found to diverge from the paraxial ray by 1.5, 10 and 32 arcsec, respectively. In comparison to the Airy disk radius of a 300 mm diameter telescope, which is ≈0.5 arcsec, these look like noticeable departures from a well-focussed image, but the quoted Airy-disk radius is an angular measure on the sky, whereas the eye peering into an eyepiece sees this multiplied by the factor m_{ang} (120) to an apparent size of 60 arcsec, and the eye's resolution is even worse at around 120 arcsec; the eye will not perceive a 32 arcsec degradation of a marginal ray from an on-axis object.

As the calculation above has been performed for an on-axis object point, the effect we are seeing is weak spherical aberration. The nature and origin of this and other aberrations will be discussed in greater detail in Sections 3.5–3.8, but for now, we note that the increasing divergence of rays at larger angles to the optical axis is what we would anticipate for the failure of the paraxial approximation as angles of incidence increase. One implication is that feeding the eyepiece with "faster" pencils of rays, such as from an $f/6$ telescope rather than our adopted $f/10$ input, will result in even higher ray angles for the marginal rays and worse performance.

We shall see now that as we shift our attention to off-axis object points, their results will deteriorate significantly. The ray paths through our trial eyepiece can be visualised using a ray trace calculated (using WinLens Basic) for $f/10$ pencils and evaluated for three object points: one located on the optical axis and two at field points 6 mm and 12 mm off axis (corresponding to 0, 7 and 14 arcmin off-axis on the sky at our adopted image scale; Figure 3.2b). Clearly, the rays for each object point pass through only a small portion of the lens, and while this arrangement appears to

(a)

Surface	ray angle θ (deg)	object distance l (m)	radius of curvature r (m)	angle of incidence sin i	RI n	RI' n'	angle of refraction sin i' (deg)	gamma γ (deg)	ray height y (m)	sag ζ (m)	ray angle θ' (deg)	image distance l' (m)	ray angle θ' (arcsec)
1st surface	0.1	-0.01798	0.02219	0.0032	1.0003	1.5168	0.0021	0.081	3.138E-05	2.219E-08	0.038	-0.04688	138.1
	1	-0.01798	0.02219	0.0316	1.0003	1.5168	0.0208	0.810	3.139E-04	2.220E-06	0.383	-0.04691	1380.2
	2	-0.01798	0.02219	0.0632	1.0003	1.5168	0.0417	1.622	6.282E-04	8.894E-06	0.766	-0.04700	2756.4
	3	-0.01798	0.02219	0.0947	1.0003	1.5168	0.0625	2.436	9.433E-04	2.006E-05	1.146	-0.04715	4124.6
lens thickness (m)		0.01830											
2nd surface	0.038	-0.06518	-0.02219	-0.0013	1.5168	1.0003	-0.0020	-0.113	4.363E-05	-4.290E-08	0.000	252.14683	0.0
	0.383	-0.06521	-0.02219	-0.0130	1.5168	1.0003	-0.0197	-1.127	4.363E-04	-4.289E-06	0.000	60.48961	-1.5
	0.766	-0.06530	-0.02219	-0.0260	1.5168	1.0003	-0.0394	-2.253	8.724E-04	-1.716E-05	-0.003	18.27913	-9.8
	1.146	-0.06545	-0.02219	-0.0390	1.5168	1.0003	-0.0591	-3.380	1.308E-03	-3.859E-05	-0.009	8.42745	-32.0

(b)

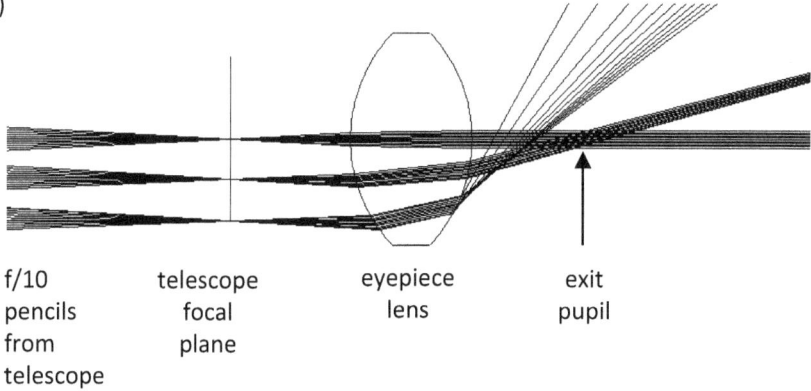

f/10 pencils from telescope   telescope focal plane   eyepiece lens   exit pupil

FIGURE 3.2 Ray paths through a single-lens eyepiece (N-BK7, r_1 = 22.19 mm, $r_2 = -r_1$, d = 18.3 mm). (a) Progressive ray trace calculation for four rays (at angles 0.1°, 1°, 2° and 3°) from an on-axis object point through each surface of the lens. (b) Three $f/10$ pencils are traced left-to-right through a 25 mm focal length single-lens eyepiece for object points located on axis, 7 and 14 arcmin off-axis (angular measure on the sky, for our adopted image scale of 1.15 arcmin per mm). The entrance pupil has a diameter of 100 mm, located 1000 mm to the left of the focal plane. The refracted rays for the on-axis object appear tolerably parallel to one another, as required for infinity adjustment, but the pencil for the 7 arcmin field point is clearly converging. The intersection of the outgoing pencils for the 0 and 7 arcmin field points indicates the location of the exit pupil, where a real image of the telescope aperture stop and entrance pupil is produced by the eyepiece. The outgoing pencil for the 14 arcmin field point is truly awful; this simplistic eyepiece obviously has a very limited usable field of view.

yield tolerably parallel rays for the on-axis object point as intended in infinity adjustment, the situation for the 14 arcmin field point is clearly much worse, exhibiting horrendous aberrations, which are discussed in Section 3.5.

However, before we abandon this trivial design completely, we should consider one potential modification. It was noted above (Equation 2.9) that the power of a thin lens is simply the sum of the two surface powers, so a lens of a given focal power can be constructed from many possible combinations of first and second surface powers. We assessed a symmetrical lens in Figure 3.2, but the rays' paths in each pencil are not symmetrical about the centre plane of the lens; they are diverging as they approach the lens from the left, and travel as a parallel pencil when they exit to the right. We might anticipate some improvement in imaging performance for the on-axis image if we "bend" the lens in such a way as to partially flatten the first surface and put more power into the second surface. Making an arbitrary decrease of the first surface power by about 20% to give $F_1 = 18.5$ D requires a revised second surface power $F_2 = 27.7$ D, and shortens the front-vertex focal length to $f_V = -16.65$ mm. A repeat of the ray-tracing exercise above improves the exit ray angles for the on-axis image to 0.3, 6 and 24 arcsec, but this is not sufficient to save the design, and other problems would be seen to worsen if we were to examine the off-axis image points. We have reached a stage of our design where the off-axis performance of the lens is poor for an $f/10$ pencil and the desired field of view cannot be attained. We must make some significant changes to the design or reduce our expectations to proceed.

The two problems we have encountered so far can be summarised thus:

- The surface powers required for a 25 mm focal-length lens are sufficiently high that the paraxial approximation is violated. We need to distribute the focal power over more than two surfaces, such as by sharing the focal power across two lenses with four surfaces. We will consider this solution in Chapters 4 and 5.

- The divergence of rays downstream of the telescope focal plane means that the lens must be much larger than the diameter of the desired field of view at the focal plane, and this limits how small the radius of curvature of the lens surface can be. We either have to sacrifice field of view to make the lens smaller, or we need some other way to limit the divergence of the rays. We explore solutions to this problem in the following section.

3.4 RELAY LENSES AND FIELD LENSES

Before detailing alternative eyepiece designs involving multiple lenses in the next chapter, we should examine one particular use of a lens as a relay lens. If we return to the fundamental paraxial equation, Equation 2.5, we see that if a thin lens is positioned in the same plane as the object so that $l = 0$, then the vergence n/l will be infinite irrespective of the power of the lens, and thus the inferred image distance must also be zero: $l' = 0$. In other words, whatever lens is placed in the object plane, the image will remain in the same location as the object: $l = l' = 0$.

You can verify this outcome for yourself by holding a magnifying glass – typically a biconvex lens – a few centimetres above the text on this page and viewing the magnified, virtual image of the words, noting also that the words seen through the magnifying glass appear to lie slightly behind the surrounding printed page, i.e., the image distance is greater than the object distance. Then place the magnifying glass down to rest on the page, and notice that the words are no longer magnified, and those seen through the magnifying glass are now the same distance away as the original print.

A lens placed coincident with a real image is called a relay lens or, in the case of eyepieces, a field lens. What is the relay lens achieving in this case? Rather than changing the *location* of the image, the relay lens refracts the rays from off-axis object points back towards the optical axis. An $f/10$ pencil of rays focussed onto a relay lens would exit the relay lens still as an $f/10$ pencil, but the principal rays for different object points would be made to converge, to a greater or lesser degree depending on the power of the field lens. This reduces the separation of pencils arising from differ-ent objects and allows subsequent lenses to be smaller than would other-wise have been the case. As we have already encountered in the preceding section, high power lenses needing small radii of curvature are inevitably restricted to a small physical radius, both practically since the radius of a lens cannot exceed the radius of curvature of its surface, and optically since the paraxial approximation is readily violated in strongly curved lenses where high angles of incidence and refraction are produced.

As a result of the challenge of designing a good quality, single-lens eyepiece, a common solution adopted in early eyepiece designs was to incorporate a field lens and then rely on a second lens further along the optical path, which became known as the eye lens since it is closer to the eye, to provide the optical power necessary to magnify the image of the sky, and to render that image in infinity adjustment.

Although the title of this chapter signifies single-lens designs, we can quickly explore the consequences for our trivial eyepiece design above if we incorporate a field lens. Although the incorporation of a field lens does not magnify or move the real image in the focal plane of the telescope, the angles that the principal rays from off-axis objects make with the optical axis *are* changed by the field lens, since the purpose of the field lens is to re-direct these off-axis pencils towards the optical axis. The benefits of this approach are apparent: not only can we reduce the size of the eye lens compared to the single-lens case, but the off-axis rays will encounter the eye lens at smaller distances from the optical axis where the angles of incidence are smaller and the failure of the paraxial condition is less acute, reducing aberrations.

Despite these advantages, placing the field lens coincident with the real image formed by the telescope optics has one very major downside, which is that any defects on the surface of the field lens, such as dust or scratches, would then be in focus with the image. The obvious solution is to move the field lens slightly away from the focal plane to avoid this, and there are just two possible directions of travel: upstream or downstream. These two choices give rise to the two eyepieces we investigate in Chapter 4, the Huygens and Ramsden eyepieces.

3.5 SEIDEL WAVEFRONT GEOMETRY

In astronomy, where stellar image sizes (the Airy disk radius and the atmospheric seeing limit) are of order 1 arcsec, there is little latitude for imperfect images, so it is important to understand in more detail what drives ray paths to behave as they do. We saw in Section 2.2 the dependence of refractive index and hence angles of refraction on wavelength, giving rise to chromatic aberration, but even for monochromatic light – light of a single wavelength – the failure of the paraxial approximation degrades imaging, as we saw in Section 3.3. The resulting imperfections are called monochromatic aberrations, and we need to understand how they arise. They depend on where the rays originate (their field angle) and which parts of a lens they pass through (their distance from the optical axis, known as the centration).

Optical designers have found it useful to split the error into different contributions that can be attributed to particular characteristics of the optical path, and to give specific names to the aberrations that arise, such as spherical aberration and coma. We examine the range of monochromatic aberrations in Sections 3.6 and 3.7, but to see how they arise, first, we need to explore wavefront and ray geometries.

As the paraxial approximation corresponds to the mathematical approximation $\sin \theta \approx \theta$ (for θ in radians; see Section 2.3), it is sensible to improve the approximation to develop a more realistic model of the passage of light through an optical system. We initiate this improvement by including the next term in the full series expansion (Equation 2.4) to make the approximation $\sin \theta \approx \theta - \theta^3/6$. Using this revision to derive new optics equations is sometimes known as third-order theory in recognition of the θ^3 term, or alternatively, Seidel theory in honour of its originator. Unfortunately, third-order theory results in a more complex calculation of image properties than under the paraxial approximation; the gain in accuracy comes at the cost of lost simplicity. Nevertheless, useful algebraic results can be obtained that guide optical designers in efforts to improve lens performance, without necessarily having to calculate large numbers of exact ray paths and then attempt to make sense of the multitude of differences that arise. The starting point of the analysis is to consider the passage of wavefronts of light, rather than rays, through the optical path.

The Spherical Wavefront

Recall that in the original presentation of the sag equation (Figure 3.2), the symbol y was used to represent the distance of the edge of the spherical optical surface from the optical axis. If y is now more generally regarded as a variable that describes the height of *any* chosen point on a spherical surface, not necessarily just the top edge, then the corresponding sag value ζ gives the axial distance to that point, i.e., measured along the optical axis from the vertex. The pair of values (ζ, y) therefore describes the axial and vertical coordinates of points on the spherical surface.

We have seen that the sine function can be written as a series expansion (Equation 2.4). The sag of a spherical surface, $\zeta = r - \sqrt{r^2 - y^2}$ (Equation 3.2), can also be written as a series expansion,[8] in this case $\zeta = \dfrac{y^2}{2r} + \dfrac{y^4}{8r^3} + ...,$ where "..." refers to higher order (smaller) terms. Spherical geometry is not restricted to a lens surface; it applies equally to a spherical wavefront. In the Seidel analysis, we approximate the wavefront by adopting just the first two terms of this expansion: $\zeta =_s \dfrac{y^2}{2r} + \dfrac{y^4}{8r^3}$, where I have added a subscript "S" to the equals sign to signify the Seidel approximation, in the same way I used "$=_p$" to signify the paraxial approximation.

The Seidel approximation for the sag describes the shape of a wavefront that is diverging or converging without any aberrations, since a perfectly spherical wavefront originates from or converges to a single (unaberrated) point. If we want to describe a not-quite-spherical wavefront, i.e., an aberrated wavefront, then we need to modify the spherical equation.

The Aberrated Wavefront

To describe an aberrated, not-quite-spherical wavefront, as a first approximation we might make the y_2 and y_4 terms of the sag expansion inconsistent with one another, for example by adding a variation to the second (y^4) term of the sag approximation given above, writing[9,10] $\zeta =_s \dfrac{y^2}{2r} + \dfrac{y^4}{8r^3} + Ay^4$, where A is a coefficient, yet to be quantified, that tells us how non-spherical the wavefront has become. This modification retains explicitly the y^4 dependence of the second term and implicitly incorporates the $1/r^3$ dependence into the constant A since, *for a given surface being analysed*, the reference spherical radius of curvature r is a constant even as the coordinate pair (ζ, y) roves over the surface.

As we want to use the aberrated sag equation to describe the shape of any general wavefront at some distance from the optical axis, it is better to use a symbol other than y, as a general point on the surface will not usually lie in the yz-plane (which incidentally we have not yet defined properly and must therefore shortly do). The distance from the optical axis to the point at which a ray meets a surface, or at which we wish to analyse the wavefront, is called the centration, and a circle about the optical axis at the same centration is referred to as a zone of the lens or wavefront. We will henceforth mostly use the symbol ρ for the centration, rather than y as first used in the sag equation, writing now[11] $\zeta =_s \dfrac{\rho^2}{2r} + \dfrac{\rho^4}{8r^3} + A\rho^4$. The final term $\Delta\zeta \equiv_s A\rho^4$ is of course the axial (horizontal) distance between the desired, spherical form of the wavefront (sometimes called the reference wavefront) and the aberrated wavefront, at centration ρ.

So far, I have avoided defining the x, y, and z axes clearly, but must now do so. The z-axis is the optical axis, and the yz-plane is the plane containing the optical axis and the principal ray for a particular object point, and therefore also containing the object and paraxial image points. The yz-plane is also called the meridional plane (or alternatively, tangential plane). The y-axis measures distances away from (above or below) the optical axis in the meridional plane, while the z-axis measures distances along the optical axis increasing from left to right. The meridional plane (yz-plane) is the plane most commonly sketched in ray diagrams, i.e., lying in the plane of the printed page, but there are of course many rays that do not lie in the meridional plane, and which are therefore not sketched in such diagrams. To complete the description of Cartesian axes, the x-axis lies in a plane at right angles to the y-axis (meridional plane) and z-axis (optical axis), so the xz-plane also contains the optical axis but is at 90° to the meridional plane, projecting in and out of the printed page in most ray-trace diagrams.

A subset of the rays within any given pencil is confined within the meridional plane for that object point. That subset, which includes the principal ray, is often referred to as a meridional (or tangential) ray fan. "Fan" is an apt description considering the rays fan out from an object point and follow paths confined to a plane.

There is another ray fan that must be introduced, called the sagittal ray fan. This subset of rays also includes the principal ray, but is at right angles to the meridional fan, and thus projects in front of and behind the meridional plane, i.e., into and out of the page of a typical ray-trace diagram. The sagittal fan for an off-axis object point is therefore inclined to the xz-plane, and in almost all instances will be refracted at the first optical surface into a new (almost-)planar surface differing from the one it initially occupied. So, while the principal ray and the meridional ray fan remain within the meridional plane and can readily be drawn on a 2D sheet of paper, the sagittal rays transition from one inclined plane to another almost-planar surface and remind us that real lenses have a 3D form, and each pencil contains some rays which move out of and back into the page.

If we analyse a wavefront as it progresses from left to right through an optical system, we may at various points along the optical axis (typically at lens surfaces) want to analyse the wavefront's shape, being interested in any departures from a spherical form. At any position along the optical axis, the centration ρ of a ray R can be regarded as having two underlying contributions, described by three variables:

- The ray R will belong to a pencil of rays admitted by the system's aperture stop (not necessarily nearby), having a principal ray P that in turn has a centration y_P at the location of interest.

- The position of the ray R relative to its principal ray P can be specified with offsets y_R and x_R, measured along the y- and x-axes, respectively.

As the x-axis is at 90° to the y-axis, we can use Pythagoras' theorem to write the centration ρ of ray R as $\rho^2 = (y_P + y_R)^2 + x_R^2$. Consequently, the aberration term $\Delta\zeta \equiv_s A\rho^4$ can be rewritten $\Delta\zeta =_s A[\rho^2]^2 = A[(y_P + y_R)^2 + x_R^2]^2$.

Further algebra allows the term in square brackets to be expanded, and terms having the same power of y_P can be grouped together, giving

$$\Delta\zeta =_s A\rho^4 = A\left(y_R^2 + x_R^2\right)^2 + Ay_P\left[4y_R\left(y_R^2 + x_R^2\right)\right] + Ay_P^2\left(6y_R^2 + 2x_R^2\right) +$$
$$Ay_P^3\left(4y_R\right) + Ay_P^4$$

We thus obtain five terms[12] which depend on the principal ray centration y_P according to the powers y_P^0, y_P^1, y_P^2, y_P^3 and y_P^4; an analysis using polar coordinates[13] rather than Cartesian coordinates results in slightly differing terms. Each term has a separate significance as an aberration, which we will interpret shortly. First, however, it is convenient to re-express the ray offsets x_R and y_R in terms of offsets measured at the aperture stop, x_A and y_A, rather than at the axial location of interest. This involves scaling all x_R and y_R values by a common factor which can be absorbed into the unspecified leading constant A, provided we allow slightly different constants A_0, A_1, A_2, A_3 and A_4 for each aberration. At the same time, we recognise that the principal ray centration y_P is related to its height in the image plane, h', again by a scaling constant that we absorb into the five As. We can therefore rewrite the wavefront aberration term as:

$$\Delta\zeta =_s A\rho^4 = A_0\left(y_A^2 + x_A^2\right)^2 + A_1h'\left[4y_A\left(y_A^2 + x_A^2\right)\right] +$$
$$A_2h'^2\left(6y_A^2 + 2x_A^2\right) + A_3h'^3\left(4y_A\right) + A_4h'^4$$

(3.3)

We can now interpret each term in this expression to determine the characteristics of the aberrations that can be recognised in this approach to the analysis. Note that at this stage, we can see in Equation 3.3 how each term in the calculation of the wavefront aberration depends on (1) the ray location within the aperture stop (x_A, y_A) and (2) the distance from the optical axis of the image point in the image plane (h'), which is obviously related to the off-axis angle of the object point, but we have *not* provided equations that show how the coefficients A_0-A_4 can be calculated; in most cases the derivations are quite involved even for a single surface,[14,15] so only the simplest results will be stated in this book, and in some cases it will be easier to assess the ray aberrations than the wavefront aberrations.

Monochromatic Aberrations

The equation we have obtained for the change in the position of the wavefront along the optical axis, $\Delta\zeta$, given by $A\rho^4$ (Equation 3.3), can be interpreted term-by-term as follows:

- The first term is independent of position in the image plane h' so is uniform for images both on-axis ($h' = 0$) and off-axis ($h' > 0$); it depends only on the ray centration at the aperture stop, via the term $y_A^2 + x_A^2$. This is called spherical aberration.

- The term linearly proportional to h' increases with distance off axis and has a directional dependence on the ray passage through the aperture stop, depending both on the ray offset in the meridional plane (y_A) and the centration $\left(y_A^2 + x_A^2\right)$. This is coma.

- The term in h'^2 is astigmatism.

- The term in h'^3 depends on the ray offset in the meridional plane (y_A) but not the sagittal direction (x_A does not feature); it corresponds to distortion.

- Finally, the term in h'^4 is independent of the ray position (x_A, y_A) within the aperture stop, so it does not differentially affect the rays within a pencil. The focussing of the rays is therefore not altered, but by changing the value of the sag (ζ changes), it shifts the position of the wavefront along the optical axis and hence shifts the distance to the point at which the rays converge. This change is evidently greater for image points further from the optical axis (i.e. at greater h'), and is known as field curvature.

We will shortly examine the Seidel aberrations in more detail, and in particular their impact on the imaging capabilities of lenses. We must emphasise beforehand, however, that the equation we have just examined for $\Delta\zeta$ tells us what the *wavefront* is doing, not what the *rays* are doing, though the two are of course linked since the rays are normal to the wavefront. A small aberration to a portion of a wavefront, advancing or retarding it by $\Delta\zeta$ relative to the ideal spherical shape, will change the slope of that region of the wavefront and therefore divert the associated ray away from the direction an unaberrated wavefront would travel. The associated ray, being the normal to the wavefront, will tilt by the same small angle that the wavefront slope changes, i.e., by $\delta\zeta/\delta y$ (where the symbol δ signifies a small change in the quantity following it). In travelling a distance l' to the image plane, the aberrated ray will therefore deviate from an unaberrated ray by a distance $l' \times (\delta\zeta/\delta y)$ in the image plane, in the y-direction. This deviation is therefore called the transverse ray aberration (TA_y) in the image[16]:

$$TA_y = l' \times \left(\delta\zeta/\delta y\right) \tag{3.4}$$

A similar equation to find the transverse aberration component in the x-direction can be written $TA_x = l' \times (\delta\zeta/\delta x)$

The longitudinal aberration (LA), on the other hand, is the separation of the principal ray and aberrated ray in the z-axis direction. The longitudinal aberration corresponding to the y-component of the transverse aberration is given in turn by

$$LA_y = TA_y \times \left(l'/y\right) \tag{3.5}$$

Note that Equation 3.4 reduces the power of y from the wavefront aberration by 1, e.g. from y^4 to y^3, and Equation 3.5 reduces it by a further power of 1, so a wavefront aberration that depends on y^4 gives rise to a transverse ray aberration TA_y that depends on y^3 and a longitudinal ray aberration that depends on y^2. This particular example will be clearer later when we see that spherical aberration, which has a wavefront error proportional to y^4, gives rise to a quadratic longitudinal ray aberration in the meridional plane.

The total wavefront shift $\Delta\zeta$ induced by multiple optical surfaces is the sum of the shifts introduced at each one, enabling wavefront aberrations to be tallied through an optical system.[17] The eventual transverse and longitudinal ray aberrations can be calculated from the total wavefront shift.

Finally, we should also note that the Seidel wavefront coefficients are usually written in the form $S_I, S_{II}, \ldots S_V$ rather than $A_0, A_1, \ldots A_4$ above, where the S-coefficients correspond to the wavefront aberrations encountered for the maximum centration and the maximum field angle, and the x_A, y_A and h' values are treated as fractional values relative to the maxima. Perhaps surprisingly, they can be calculated from *paraxial* ray-trace data for just two rays: a marginal ray for an on-axis object point, and a principal ray at the maximum desired field angle.[18,19] We won't derive or explicitly calculate most of the A or S values in this book, but if you read about them elsewhere,[20] or inspect values reported by ray-tracing programs such as

WinLens Basic as we will do in later chapters, it is typically the S-forms expressed in units of millimetres that you will encounter. Our coefficients A_0, A_1, A_2, A_4 and A_3 feed into S_I, S_{II}, S_{III}, S_{IV} and S_V, respectively; note the reordering of A_4 and A_3 in this list, which occurs because A_2 and A_4 have related impacts on image formation and are therefore treated as consecutive Seidel terms S_{III} and S_{IV}.

We can now seek to gain intuitive as well as mathematical insights into how the five Seidel aberrations can be influenced and ideally controlled. We should note that as the wavefront aberration term $A\rho^4$ models the shift in wavefront position, which has units of length (e.g. mm), and the centration ρ also has units of length, then A must have units of 1/length^3, e.g. mm^{-3}. This reminds us, as noted above, that A has an implicit $1/r^3$ dependence. With the exception of field curvature, the Seidel aberration coefficients therefore have a cubic dependence on surface curvature[21] that embeds many non-linear dependences on surface power, meaning the aberration can change sign and strength in complicated ways as the radius of curvature of a surface is varied. Moreover, a change in the radius of curvature at one surface affects the wavefront reaching the next surface and therefore changes the ray paths at subsequent surfaces too. These two features – the cubic dependence of wavefront aberrations on radius of curvature and the compounding of changes at multiple optical surfaces – highlight why aberration control in optical design is a non-linear problem, and why the experience and insight of expert designers have been key to solving optical design challenges. Computer optimisation routines have aided the designer in that process in recent decades, but have not hitherto replaced the skill of the craftsperson. Whether artificial intelligence (AI) systems do so in the future, or perhaps more realistically, how long it will take until they do, remains to be seen.

3.6 SPHERICAL ABERRATION

Spherical Aberration of a Surface

The wavefront error associated with spherical aberration for a refracting spherical surface can be derived with a page or so of algebra[22] describing its y^4 dependence, and then the transverse aberration (proportional to y^3) and longitudinal aberration (proportional to y^2) can be inferred by applying Equations 3.4 and 3.5. An alternative approach is to derive the longitudinal ray aberration directly via a different route, revising the derivation of the fundamental paraxial equation (Equation 2.5) with improved approximations. If we retain smaller contributions to the ray pathlength, such as the quantity ζ in the sum $l + \zeta$ (see Figure 2.5), along with other "small" differences that were ignored in the paraxial derivation, we still require a page or so of algebra, but usefully obtain the revised equation[23]

$$\frac{n'}{l'} = \frac{n}{l} + F_{surf} + y^2\left(\frac{n'}{2l'}\left(\frac{1}{r} - \frac{1}{l'}\right)^2 - \frac{n}{2l}\left(\frac{1}{r} - \frac{1}{l}\right)^2\right) \tag{3.6}$$

In comparison with the fundamental paraxial equation for a surface (Equation 2.5), we can see that an additional term depending on y^2 has appeared in this more detailed treatment. The fundamental paraxial equation, on the other hand, did not depend on the centration y at all, though our exact ray tracing calculations in Section 3.3 indicated otherwise, since rays launched at greater angles to the optical axis deviated from the paraxial ray.

If we relabel l' in the fundamental paraxial equation (Equation 2.5) as l'_p, and subtract Equation 2.5 from Equation 3.6, we can obtain the longitudinal aberration directly as $l' - l'_p \approx \left(\frac{l'^2_p}{n'}\right)y^2\left(\frac{n'}{2l'}\left(\frac{1}{r} - \frac{1}{l'}\right)^2 - \frac{n}{2l}\left(\frac{1}{r} - \frac{1}{l}\right)^2\right)$.

We can make more intuitive sense of it if we restrict our attention to rays coming from an object at a great distance, for which $l = -\infty$, as then the term $(n/2l)$ is zero. Moreover, we know that in this case, the image will be formed close to the second focal point of the surface, i.e., $l' \approx f' = n'/F = n'r/(n' - n)$, and for an air-to-glass surface where $n' \approx 1.5$, this simplifies to $l' \approx 3r$. We can therefore replace l' with $3r$ in the remaining part of the y^2 term to express the longitudinal spherical aberration of an air-to-glass surface with the object at infinity, as $l' - l'_p \approx -\left(\frac{l'^2_p}{n}\right)y^2\left(\frac{n'}{6r}\left(\frac{1}{r} - \frac{1}{3r}\right)^2\right) \approx -\frac{2}{27}\frac{l'^2_p y^2}{r^3}$.

Although this expression applies only to the special case of an object at infinity for an air-to-glass surface, it contains important general lessons:

- Longitudinal spherical aberration depends on y^2, so it deteriorates quickly with increasing centration, i.e., for rays passing through the outer zones of the optical surface.

- If r is positive, then the marginal rays cross the axis closer to the surface than the paraxial rays.

- It varies inversely with r^3, so it gets rapidly worse for surfaces with a small radius of curvature, i.e., at high focal power.

- *Crucially,* since the power of *r* is odd (3), the longitudinal spherical aberration depends on the *sign* of *r.* This offers lens designers the hope that they can offset a positive value of the aberration incurred at one surface by a counterbalancing negative term at another surface. Without this sensitivity to the sign of the surface radius of curvature, there would be little hope of controlling this aberration.

Note that this aberration term was obtained for an object point on the optical axis, but the wavefront analysis showed that spherical aberration is the same for off-axis points. Conceptually, as the optical axis can be rotated to any off-axis object point without changing the geometry of the analysis, and without changing the spherical surface, this outcome is expected. Precisely because it affects image points on the optical axis, it is one of the major aberrations to be controlled in optical design. All other monochromatic aberrations vanish on the optical axis, so if a limited field of view can be tolerated, then spherical aberration is often the most important monochromatic aberration to correct at the outset. (The classical Schmidt camera perfectly illustrates the *opposite* strategy; for that instrument, *maximum* field of view is the goal, and this is achieved by accepting spherical aberration uniformly across the field with the benefit that all the off-axis aberrations are eliminated.)

Spherical Aberration of a Lens

The spherical aberration equation for a single surface (Equation 3.6) can be extended to a pair of surfaces, i.e., a lens. For incoming parallel rays being brought to a focus by a lens, spherical aberration can be described as a failure of the lens to bring marginal rays to the same focus as the paraxial rays. For a positive lens, the peripheral rays will commonly focus closer to the lens, as shown in Figure 3.3a, which is described as positive spherical aberration (even though $l'_M - l'_P < 0$). For a negative lens, the opposite happens, but it is important to note from Equation 3.6 that the effect depends on the object and image distances, i.e., how the optical surface is used, so it would be a mistake to over-generalise outcomes.

The spherical aberration can be quantified in a number of ways. One way as introduced above is to state the difference in the distances at which a marginal ray crosses the optical axis compared to the paraxial ray, which in the case of Figure 3.3a would be 15 mm. Alternatively, since the image distances appear in the fundamental paraxial equation for a lens in the form $1/l'$, which you may recall from Chapter 2 is known as a vergence,

(a)

(b)

-25mm   F′

(c)

Scale
10.0mm

Spacing
4.0mm

Def: -8.0mm

-16.0mm   -12.0mm   -8.0mm   -4.0mm   .0mm

on-axis

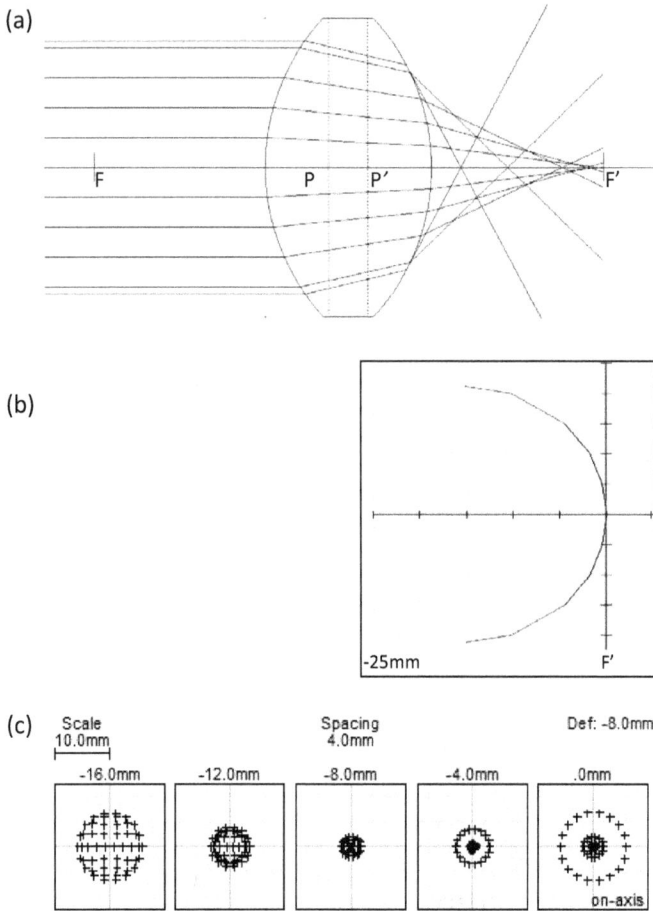

FIGURE 3.3   Spherical aberration arising from a positive lens (r_1 = +22.39 mm, r_2 = −22.39 mm, d = 17.6 mm, aperture = 30.8 mm, N-BK7, $f′$ = 25 mm @ 587.6 nm). (a) Ray trace of positive lens for an object at infinity (far to the left). The locations of the first and second focal points F and F′, and the first and second principal planes P and P′, are indicated. (b) Longitudinal aberration plot showing (horizontal axis) the distance along the optical axis at which rays that struck the lens at different heights (vertical axis) above or below the optical axis cross the optical axis. Paraxial rays strike the lens just off the optical axis and therefore correspond to points at far right, while rays closer to the edge of the lens cross the optical axis further to the left. The quadratic (y^2) dependence of spherical aberration on the height of the rays at the lens (y) is evident from the resulting curve being a parabola lying on its side. (c) Spot diagrams corresponding to an on-axis object point, showing the distribution of a set of rays at 20%, 40%, 60%, 80% and 84% of the radius of the aperture when they intersect five planes close to the paraxial focus of the lens. The five planes are at longitudinal distances −16, −12, −8, −4 and 0 mm from the paraxial focus. Image scale as shown.

the spherical aberration can instead be defined as the difference in the vergences of the paraxial (P) and marginal (M) rays, $1/l'_M - 1/l'_P$, with the result being in dioptres (D); this approach is also valid, but is not what we employ in this book.

Two ways of visualising spherical aberration, and other aberrations, are shown in Figure 3.3b and c: a longitudinal aberration plot and spot diagrams. The horizontal axis of a longitudinal aberration plot represents the distances along the optical axis at which various rays cross the principal ray. The vertical axis indicates the centration y at which each ray passes through the aperture stop (which in Figure 3.3 coincides with the lens); see also Figures 2.5 and 3.1. From the y^2 dependence of the longitudinal spherical aberration term in Equation 3.6, it comes as no surprise that the longitudinal aberration plot for spherical aberration takes the shape of a parabola lying on its side, symmetric about $y = 0$ (the optical axis). The curve illustrates how the aberration deteriorates as rays strike the lens at greater distances from the principal ray. In the absence of aberrations, the curve would be a vertical line indicating that rays of all centrations cross the principal ray at the same axial position.

Spot diagrams (Figure 3.3c) are produced by populating the aperture stop with a large number of rays, usually distributed in rings, and showing where they pass through a set of closely spaced inspection planes at right angles to the optical axis, near the image plane. Typically, five inspection planes might be shown, to illustrate how well the rays are focussed at slightly different distances along the optical axis. The default location for the central inspection plane of the set is usually the paraxial image, but the inspection plane at the paraxial image does not necessarily coincide with the sharpest focus of an aberrated pencil. For an on-axis object point affected by spherical aberration, as in Figure 3.3, the best focus is somewhere between the paraxial focus and the marginal ray focus. A location sometimes regarded as a reasonable focus compromise coincides with the narrowest waist of the hourglass-shaped envelope of rays that exists between the marginal-ray focus and the paraxial focus, and the image obtained here is called the "circle of least confusion". It is evident in Figure 3.3c that the circle of least confusion provides a tighter concentration of rays than the paraxial image plane. Offsetting the central inspection plane from the paraxial image to the preferred image plane is called "defocus".

The scale at which spot diagrams are plotted can be varied to suit the system. Spot diagrams can also be created for object points away from the optical axis, to indicate how the aberrations vary with field angle, as we will see in later Figures.

Spot diagrams therefore render an extremely helpful visual impression of the image quality being achieved by an optical system. They can be further quantified if desired, such as by calculating the radius of a circle enclosing 80% of the rays. However, despite the excellent visual representation of overall aberrations that spot diagrams provide, they do not tell the designer where in a system the various aberrations arise, nor the fractional contribution of each one; it may be that only the most obvious one is evident in the spot diagrams. To determine more detailed and useful information, the designer needs to review the Seidel coefficients S_I–S_V at each surface, aided by the longitudinal aberration plots, spot diagrams and other aberration plots that are introduced later.

We note that the spherical aberration of a surface (Equation 3.6) depends on the object and image distances and the radius of curvature. As the object distance for the *second* surface of a lens is the image distance produced by the first surface, not the object distance of the first surface, it is not surprising that the spherical aberration contributions of the individual surfaces of a symmetric biconvex lens are not generally the same.[24] Consequently, the total spherical aberration of a lens can be changed by shifting power from one surface to the other, i.e., by bending the lens to retain the same overall power and focal length but redistributing the power more advantageously across the two surfaces. We demonstrated this briefly in Section 3.3 using exact ray tracing equations to examine the deviation of rays away from the paraxial ray for different lens shapes. In the light of preceding comments, it should not come as a surprise that for a thin lens used to produce a nearby image of a distant object, the lens shape that minimises spherical aberration is not symmetrical. A clue to the more desirable shape can be obtained by considering the arrival of incoming parallel rays at the first surface of a plano-convex lens. If the first surface is plano, then the parallel rays will undergo no refraction at that surface, and all of the refraction must occur at the second, convex, surface. This requires that the desired angles of deviation be achieved in one refraction. Switching the lens around does not change its effective focal length, but rays arriving at the first surface will now encounter a convex face, will be refracted slightly towards the optical axis, and will then arrive at the plano surface already inclined to the normal. The plano surface will therefore also refract the rays towards the optical axis, in this way producing the required deviation via two smaller refractions instead of one large refraction. Much less spherical aberration is induced by two small refractions compared to one large refraction, and so the convex-first orientation is preferable in this instance.

The effect of an optical surface or lens is to convey the rays from an object point to an image point. Due to the reversibility of light, we can equivalently follow the path of light forward and backward through an optical system, so the distinction between the object and the image is erased. In optical systems, where the image formed by one surface becomes the object for the next, the distinction is further blurred. It becomes more useful to speak of "conjugates"; two points are conjugates if one is the optical image of the other, and we don't need to identify which is which. It can be very helpful therefore to describe the imaging properties of lenses in terms of conjugates, without distinguishing between object and image, which may depend on the use to which the lens is put. Our discussion of the spherical aberration of a plano-convex lens is a case in point. The conclusion from the analysis above is that if a plano-convex lens is to be used in a pencil of rays with one nearby conjugate and one distant conjugate, the plano side should face the nearby conjugate. If, on the other hand, the two conjugates are symmetric about the lens and nearby, then the plano-convex lens should either be set aside and replaced by a symmetric biconvex lens, or two plano-convex lenses should be used back-to-back, with the convex sides facing one another. The latter strategy has the added benefit of sharing the refraction across four surfaces rather than two, thus reducing the aberrations introduced at each stage. If bending lenses is the first tool in the lens designer's bag of tricks to improve optical performance, then sharing powers across more surfaces is the second. We will need to utilise all of these tricks, or should I say strategies, in the design of better eyepieces in later chapters.

It was noted above that for a plano-convex lens used to focus the light from a distant object (conjugate) to a nearby image (conjugate), a convex-first orientation was better than plano-first, but how does a convex-first plano-convex lens compare to a symmetric biconvex lens? A detailed algebraic treatment[25,26] for a thin lens shows that the best shape is intermediate between the two, possessing a biconvex form with the second surface weaker than the first. The algebra is not so neat, and the longitudinal spherical aberration of a thin lens is perhaps best written

as $\quad l' - l'_p \approx -l'_p \dfrac{y^2}{8f'^3} \dfrac{1}{n(n-1)} \left(\dfrac{n+2}{n-1} q^2 + 4(n+1)pq + (3n+2)(n-1)p^2 + \dfrac{n^3}{n-1} \right)$

where we have introduced two new parameters p and q which are called the Coddington position and shape factors, respectively. These are defined as $p \equiv \dfrac{l'+l}{l'-l}$ and $q \equiv \dfrac{r_2+r_1}{r_2-r_1}$. For example, an object distance $l = -\text{infinity}$ will give position factor $p = -1$, while the symmetric case $l = -2f'$ with $l' = +2f'$

gives $p = 0$. As for lens shapes, a symmetric lens with $r_2 = -r_1$ has shape factor $q = 0$, while plano-convex lenses have shapes $q = +1$ in the convex-first "(|" orientation and $q = -1$ in the plano-first "|)" orientation. Meniscus lenses have $q < -1$ or $q > 1$.

The spherical aberration of a lens employed for a particular purpose (i.e. with l_p and f' specified) can be minimised by minimising the term in the brackets, but as the encompassed terms in q^2, p^2 and n^3 are all positive, the only way of generating a negative contribution to reduce the sum is by having p and q of opposite sign so pq is negative. Not surprisingly, then, given the preceding discussion, the spherical aberration of a thin lens is minimised by suiting the lens shape (q) to the object position (p). Moreover, the four terms in the large brackets cannot achieve a sum of zero; it can at best be minimised but not eliminated in a single thin lens.

The lens shape which minimises the spherical aberration is described by $q = -\dfrac{2(n^2 - 1)}{n + 2}p$, and for $n_g = 1.5$, this corresponds to $q = -(5/7)p$. For the specific case of an infinite (distant) conjugate on side 1 of the lens and the finite (near) conjugate on side 2, $p = -1$, and hence we want $q = 5/7$. This is achieved with $r_2 = -6r_1$, which is almost plano-convex in a convex-first "(|" form, which provides an algebraic explanation of the lens bending results described above. (For $n_g = 1.67$, the corresponding relationships are $q = -(32/33)p$, $q = 32/33$, and $r_2 = -65r_1$, which is even more plano-convex-like.)

Four cases are illustrated in Figure 3.4 for a 20 D lens (i.e. $f'_E = 50$ mm) for an object at infinity: plano-convex orientated plano-first; symmetric biconvex; best-form; and plano-convex orientated convex first. Following the design process above (Section 3.2), a thin-lens symmetric design implies $F_1 = F_2 = 10$ D, and for $n_g = 1.5$, we infer $r_1 = 50$ mm. For a physical radius of 17.4 mm, this gives a sag $\zeta = 3.1$ mm, so we impose a thickness $d = 7.2$ mm. Refining the surface curvatures for a thick lens using N-BK7 glass ($n_g = 1.5168$ at 587.6 nm), we expect to need:

- $r_2 = -25.83$ mm for the plano-convex lens (plano-first)

- $r_1 = 50.39$ mm and $r_2 = -r_1$ for the symmetric lens

- adopting $F_1 = 6F_2$ for the best-form lens, $r_1 = 29.77$ mm and $r_2 = -178.7$ mm

- $r_1 = 25.83$ mm for the plano-convex (convex-first).

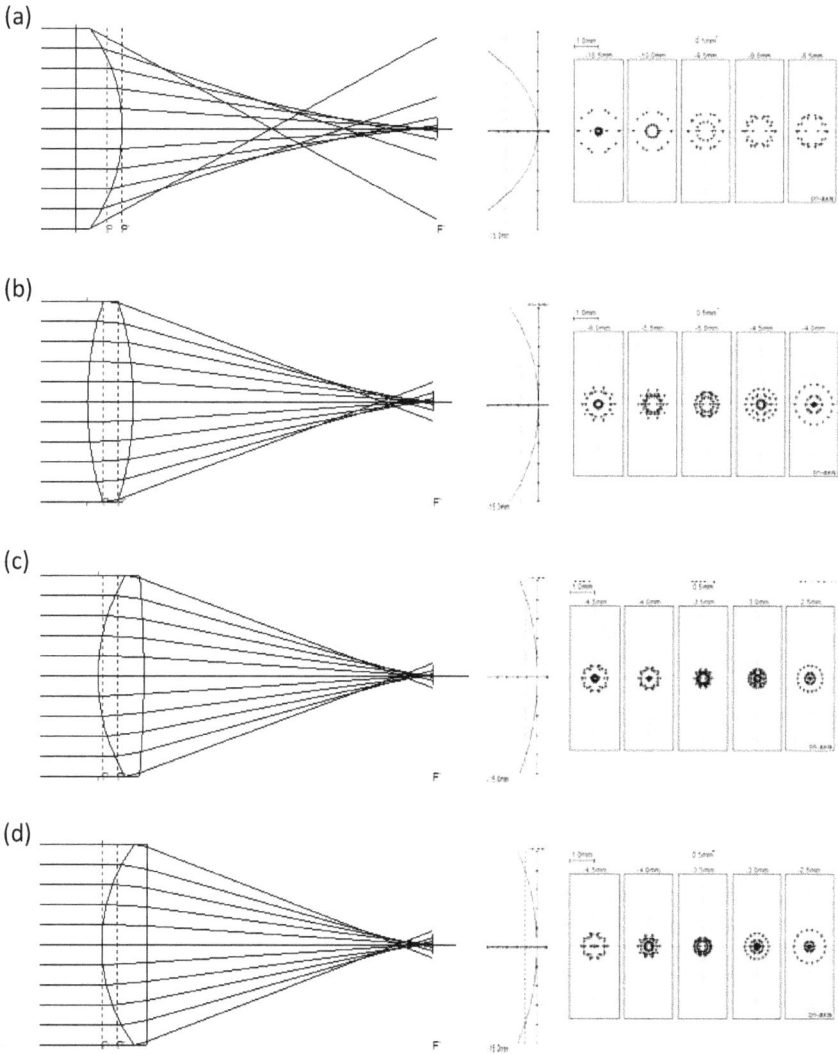

FIGURE 3.4 Bending a lens alters its spherical aberration. Ray traces, longitudinal aberration curves and spot diagrams, for $f' = 50$ mm lenses (N-BK7, $d = 7.2$ mm) and on-axis object at infinity (to far left). (a) plano-convex lens with convex side facing near conjugate (poor choice), (b) symmetric biconvex lens, (c) best-form lens, (d) plano-convex lens with plano side facing near conjugate (good choice). The locations of the second focal point F′ and the principal planes P and P′ are indicated for each lens. Note that the lens position along the optical axis is slightly different in each case, and the locations of the principal planes relative to the lens surfaces shift as the lens is bent, but the distance between the second principal plane P′ and second focal point F′ is the same in all cases, being the effective focal length of the lens (50 mm).

It is evident from the diagram that the best-form lens and the plano-convex lens with the plano side facing the near conjugate have almost identical performance, so the manufacturing simplicity of the plano-convex lens compared to a best-form lens makes it a sensible choice in many instances.

For completeness, we should note that in applying the third-order/Seidel approximation, we have ignored higher-order terms, and consequently, we may fool ourselves into thinking we have a lens or system optimised to minimise aberrations, only to discover it doesn't work as well as intended. Just as the paraxial approximation failed for rays passing through the lens further from its centre, so does the Seidel approximation if the ray distances (or surface curvatures) are large enough. In this case, the next order of correction to consider is fifth order (see Equation 2.4). In this book, we shall not cover fifth-order details, other than to note their existence[27] and the necessity ultimately of considering them if an aberration calculation and optimisation is to be more complete. Fifth-order spherical aberration is a case in point, and the ability to compensate spherical aberration in one component by the spherical aberration in another is partly undermined by the existence of fifth-order terms; full correction is limited to one zone of the aperture.

3.7 OFF-AXIS MONOCHROMATIC ABERRATIONS

The terms in Equation 3.3 that explicitly depend on h', the principal ray height in the image plane, are collectively called off-axis aberrations because they do not exist on the optical axis where $h' = 0$, and they increase in importance the further off-axis an image point is. It is the off-axis aberrations that usually limit the useable field of view of an eyepiece. The rays from an on-axis object point pass symmetrically through an optical system, with rays launched below the optical axis mirroring those above the optical axis, and rays in the sagittal fan having equivalent rays in the meridional plane. However, this is not the case for object points lying off the optical axis, where meridional ray paths below the optical axis are different from those above the optical axis, and both are different from ray paths in the sagittal fan. It is this asymmetry of the ray paths that leads to the off-axis aberrations, which we detail now.

Coma

The pencil of rays launched from an object point below the optical axis will include the principal ray and numerous others. Because of the asymmetry described above, meridional rays passing through the upper portion of a single, positive lens may intersect the principal ray at greater or lesser distances along the optical axis than rays with the same centration travelling through the lower half of the lens. Another way of expressing this is that the upper and lower marginal rays intersect in a different location than the principal ray and its adjacent rays. This aberration is called coma. As the lateral magnification of a lens or system is defined as the ratio of the image height to the object height, and we now discover that the image height can depend on the zone of the lens through which the rays passed, coma can also be described as a variation of lateral magnification with lens zone or (equivalently) centration.

If we inspect the distribution of rays from a particular zone of the lens in a transverse plane near the paraxial image point, we will find that they are distributed in a circular pattern offset from the principal ray, displaced either towards or away from the optical axis. The displacement and diameter of the circular distribution of rays in the image plane both increase with centration, resulting in a conical, comet-like image, which gives this aberration the name coma.

Rays can be traced using the exact ray-tracing formulae (Equation 2.3), but applying the Seidel theory helpfully yields an expression for the length and width of the comatic image.[28] Its length in the meridional plane is 1.5× its width along the x-direction, so it points directly towards or away from the optical axis (Figure 3.5). The dimensions depend on the image height of the principal ray h' (it increases in size further off-axis, in a linear fashion, and changes direction below the optical axis), the square of the marginal ray centration p, the cube of the focal length f' of the lens (and is thus sensitive to the sign of f'), the refractive index of the glass, and an additional term which depends on the distance of the object and the shape of the lens. This additional term is key to controlling coma and may be written[29,30,31,32] thus:

$$\frac{3(2n+1)}{4n}p+\frac{3(n+1)}{4n(n-1)}q \tag{3.7}$$

where p and q are the Coddington position and shape factors, which we met above. As p and q can take positive and negative values, the coma term in Equation 3.7 can be forced to zero, thus eliminating coma by a choice of lens shape appropriate to the object location, i.e., by requiring $q=-(2n+1)\left(\dfrac{n-1}{n+1}\right)p$. For two representative refractive indices $n = 1.50$ and $n = 1.67$, the requirement for zero coma reduces to $q = -0.8p$ and

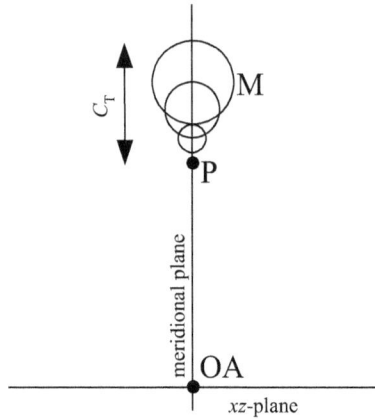

FIGURE 3.5   Comatic image in the image plane (schematic). The meridional (yz) and xz planes intersect along the optical axis (OA), while P indicates the point at which the principal ray for a particular off-axis object point pierces the image plane and where the paraxial image would be expected. Coma results when the rays passing through different zones of a lens, i.e., at different centrations, are focussed away from the principal ray. The rays passing through each zone populate a circle in the image plane at distances that increase with centration, and whose circular radius also increases with centration. The comatic circles for three zones of the lens are illustrated, with the largest circle corresponding to marginal rays (M). The maximum extent of the comatic image is labelled C_T.

$q = -1.0p$, respectively. For a positive lens used to image an object at infinity ($p = -1$), the lens shape for zero coma is $q = 0.8$ (for $n = 1.5$), i.e., close to the plano-convex form "(|" with the plano side facing the nearby conjugate. Fortuitously, this is close to the shape that minimises spherical aberration. A lens that has been designed to minimise both spherical aberration and coma is often called aplanatic or "best-form".

For other object distances, and hence different position factors p, different lens shapes q will be required to eliminate coma. As the object distance diminishes from infinity, the coma-free lens shape becomes gradually more symmetric, reaching perfect symmetry for the case where $l' = -l = 2f'$, i.e., for $p = 0$, which requires a shape factor $q = 0$, i.e., a symmetric lens. This object and image combination is known as a "$4f$" configuration, since for a thin lens the object and image are then 4 focal lengths apart, and the lateral magnification is -1, the minus sign signifying an inverted image. (In a thick lens $4f$ configuration, the object and image are separated by $4f' + \overline{PP'}$, where $\overline{PP'}$ is the hiatus of the lens.)

In seeking to demonstrate coma for a spherical lens, we face the challenge that coma is not usually seen in isolation from other aberrations, but rather alongside them. Figure 3.6a and b shows ray traces and spot diagrams for two lenses having almost identical spherical aberration but opposite coma, to emphasise the differences in the ray paths for these two contrasting comatic cases and to illustrate the effect on the image produced (spot diagrams). Figure 3.6c shows the lens shape that delivers minimal spherical aberration and zero coma. The table in the inset shows the Seidel coefficients S_I for spherical aberration and S_{II} for coma at each surface in Figure 3.6c, in millimetres, where it can be seen that both surfaces introduce coma, but of opposite sign, so the contributions from the two surfaces cancel one another out leaving the lens essentially coma-free overall.

In the design process, the presence of coma is often tracked by inspecting the "offence against the sine condition", or OSC. The sine condition is an alternative way of expressing the design requirement for zero coma, specifically that the ratio $\sin\theta/\sin\theta'$ must be constant for all ray paths from a particular object point to its image if the rays are to converge at a point, and it applies to a single optical surface, a lens or a complex optical system.[33,34] The value of the constant is the lateral magnification of the lens (multiplied in the cases of a single surface or imaging in a medium other than air by n'/n) which echoes the description of coma as a variation of lateral magnification with ray centration. Checking whether this condition is in fact met for a particular ray trace indicates whether the coma-free condition is met; deviation from (offences against) the sine condition indicates coma and ideally would spur efforts to control it better.

Astigmatism and Field Curvature

In the Seidel aberration series (Equation 3.3), spherical aberration is independent of image height h' while coma increases linearly with h'. The subsequent Seidel term, which we call astigmatism,[35] increases as h'^2, so this term will grow faster than coma at large field angles, and may overtake coma in significance, especially if coma is controlled and minimised. We therefore need to understand how astigmatism affects the image and how, if at all, it can be controlled.

It is obvious from the Seidel wavefront term $A_2 h'^2 \left(6 y_A^2 + 2 x_A^2\right)$ that the x_A and y_A variables affect the wavefront error differently, so rays in the meridional plane will behave differently to rays in the sagittal fan. The effect of this term is to displace the image point of sagittal rays from the ideal non-astigmatic image point by a certain distance along the optical axis,

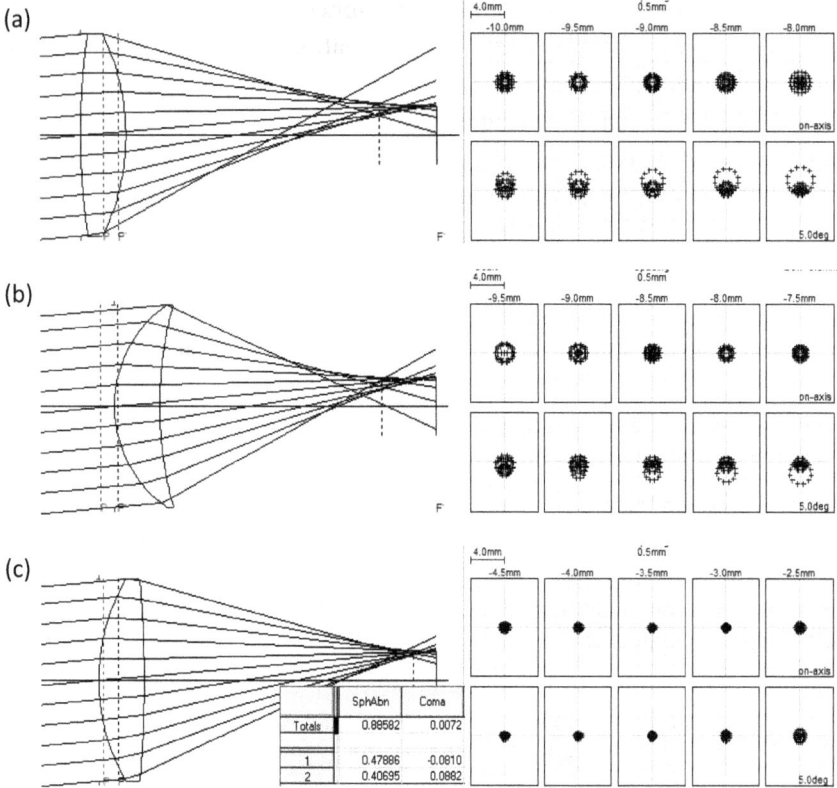

FIGURE 3.6 Coma and spherical aberration for three positive lenses. All three lens have $f'_E = 50$ mm, use N-BK7 glass of thickness 7.2 mm, and have rays shown for an object at infinity, 5° off axis. All ray plots and all spot diagrams are to the same scale. Different defocus values have been adopted as a consequence of the differing ray paths. Cases (a) [$r_1 = 15.28$ mm, $r_2 = -49.97$ mm] and (b) [$r_1 = 18.36$ mm, $r_2 = 55.06$ mm] have essentially the same spherical aberration as seen in the on-axis spot diagrams (upper rows), but opposite coma, evident from the contrasting ray paths for the marginal rays, and the opposite orientation of the comatic flare in the off-axis spot diagrams (lower rows). When the marginal rays are focussed further from the optical axis , the coma is positive (case (a)), while in negative coma (case (b)), marginal rays are focussed closer to the optical axis than the principal ray. Case(c) [$r_1 = 29.77$ mm, $r_2 = -178.7$ mm] is the best form for this lens, showing that both spherical aberration and coma have been greatly reduced as a result of bending the lens appropriately for the intended object distance. The first two Seidel aberrations (spherical aberration S_I and coma S_{II}) incurred at each surface are tabulated in the inset (mm), where it can be seen that coma is introduced in the first surface but is balanced out at the second surface.

and for the image point of meridional rays (also called tangential rays) to be displaced by three times that value (ratio 6/2). Clearly, then, the image points formed by sagittal and tangential ray fans are unhelpfully split. As only the tangential rays are focussed at the tangential-ray image distance and, for a positive thin lens, the sagittal rays are still converging, the tangential image "point" is in fact an extended line lying at right angles to the y-axis (meridional plane). Similarly, at the image distance for sagittal rays, tangential rays will have begun to diverge, and the sagittal image will similarly constitute a line, in this case aligned along the y-axis. In between these two positions, the image of the object point will consist of a blurred circle. There therefore exist two image surfaces, one for meridional rays and one for sagittal rays, which curve away from the non-astigmatic image surface proportional to the square of the off-axis image distance, h'^2, meaning the tangential- and sagittal-fan image surfaces are parabolas (or more correctly, 3D paraboloids).

The separation of the sagittal and tangential astigmatic image surfaces is, unhelpfully, largely independent of the lens shape and thus immune to attempted correction by bending.[36] To illustrate this, the astigmatism of the three lenses having $f' = 50$ mm introduced in Figure 3.6 can be inspected. Longitudinal ray plots are given in Figure 3.7 for object points on-axis and 5° off-axis, and they are shown separately for meridional (tangential) and sagittal ray fans. Ray traces are drawn only for the off-axis object point to simplify the diagram. Spherical aberration is immediately evident from the sideways parabola in the longitudinal aberration diagrams, and coma is immediately evident from the asymmetric cross-over of the marginal rays for the off-axis object point in Figure 3.7a and b and the vertical offset of the spherical aberration parabolas for meridional rays in those two longitudinal aberration diagrams, but for now we want to focus on astigmatism. For the on-axis object, the meridional (tangential) and sagittal rays share the same symmetry about the optical axis, and therefore, they cross the optical axis at the same, paraxial focal point. For the off-axis object point however, the principal ray and low-centration rays intersect around 0.5 mm to the left of the on-axis paraxial focus (this is field curvature, which we comment on below), and crucially the meridional (tangential) rays focus around 0.2–0.3 mm to the left of the sagittal rays; *this* difference is astigmatism. The Seidel S_{III} (wavefront) values are given in the inset tables, and are 0.024 mm for Figure 3.6a, 0.033 mm for the best-form lens (c), and 0.038 mm for the meniscus lens (b); clearly there is some small change of astigmatism upon bending our real lens, but not

(a)

	Astig	PtzCv
Totals	0.0235	0.0244
1	0.0040	0.0061
2	0.0195	0.0183

(b)

	Astig	PtzCv
Totals	0.0379	0.0225
1	0.0222	0.0337
2	0.0156	-0.0112

(c)

	Astig	PtzCv
Totals	0.0328	0.0242
1	0.0137	0.0208
2	0.0191	0.0035

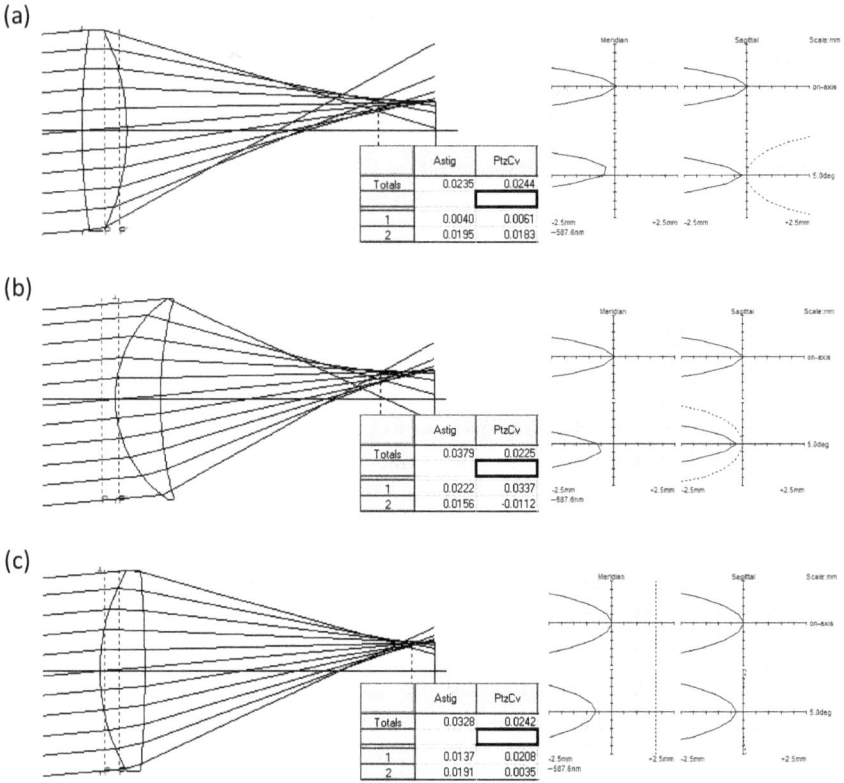

FIGURE 3.7 Astigmatism and Petzval curvature of three positive lenses. Three lenses with $f' = 50$ mm, as in Figure 3.6, tabulating the Seidel coefficients (in mm) for astigmatism (S_{III}) and Petzval curvature (S_{IV}), and showing longitudinal ray plots for on-axis (upper) and 5° off-axis (lower) object points, separately for meridional (=tangential; left) and sagittal (right) ray fans. The different axial locations of the focal points for the off-axis tangential and sagittal ray fans illustrate the presence of astigmatism, as also seen in the non-zero S_{III} values. The three different lens bendings evident in the Figure have made only modest differences to the astigmatism.

sufficient to eliminate the aberration. For a *thin* lens that coincides with the aperture stop, S_{III} depends[37] quadratically on incident ray parameters as $(\theta h)^2$, and linearly with lens power F, but bending a thin lens doesn't alter either factor. In other words, while you can bend a thin lens to eliminate coma and substantially minimise spherical aberration, you cannot bend a thin lens to eliminate astigmatism. Admittedly, real lenses are not thin lenses, and the aperture stop cannot coincide with all lenses in a system, so we can find some sensitivity of astigmatism to bending of real lenses, but we should not be surprised that its impact is often limited.

It may be noted that the term $(\theta h)^2$ is always positive, so the sign of the astigmatism of a thin lens depends on its power F. We can potentially balance astigmatism between two or more positive $(F > 0)$ and negative $(F < 0)$ thin lenses, but *not* if they are in contact with one another in a doublet, as the equation for the power of a separated thin lens pair $F_E = F_1 + F_2 - dF_1F_2$ (Equation 2.17) reduces to the simple power sum $F_1 + F_2$ if the separation $d = 0$. Therefore, two thin lenses in contact revert to the same total power as their sum, and thus the astigmatism of the thin-lens doublet also depends just on their sum, which will remain net positive if the lens is positive. Trading off the astigmatism of positive and negative thin lenses can therefore only be accomplished if they are separated from one another. We will find later that we need other free parameters to be able to reduce astigmatism, and this requires the introduction of a second lens into the eyepiece design; this is an additional limit on the usefulness of a single-lens eyepiece. Together, the immunity of astigmatism to lens bending and its high dependence on off-axis distance and power foreshadow the difficulty lens designers face in controlling this aberration in eyepieces which are, by their nature, high-positive-powered optics.

Our discussion of astigmatism would be seriously incomplete without a discussion of the next term in the Seidel expansion, S_{IV}, which originates with the field curvature term A_4 in Equation 3.3, viz., $A_4h'^4$. The absence of any dependence on x_A or y_A in the A_4 term means that the entire pencil of rays associated with the off-axis image point at h' is affected similarly by the change in wavefront sag, which has the effect of changing the distance along the principal ray to the point of convergence, thus moving the image surface for this field point forwards or backwards relative to the on-axis image plane, but the ray intersection and hence image quality remains intact for all rays in that pencil. Field curvature therefore bends the focal surface, so focus is attained at different distances along the optical axis for object points further off-axis, but the point-like focus of their pencils is not disrupted.

It is easier to develop an intuitive understanding of the change in image distance we call field curvature by considering the behaviour of paraxial rays, as we did for spherical aberration. If two object points O_1 and O_2 lie in a plane perpendicular to the optical axis, with O_1 on the optical axis and O_2 off-axis, then the distance a ray must travel from O_2 to the vertex of a surface or lens is greater than the distance from O_1. The fundamental paraxial equation for a thin lens (Equation 2.7) allows us to calculate the distance to the images O_1' and O_2': the further an object is from a positive surface or lens, the closer the image will be. (Recall that the Cartesian distance of an object point to the left of the surface or lens is negative.) As O_2 is further from the vertex than O_1, the off-axis image O_2' is therefore closer to the lens than the on-axis image O_1', which requires that the focal surface curves towards the lens as the off-axis image height h' increases. (For a negative-power surface or lens, the opposite curvature results.) This behaviour is called field curvature and is unavoidable for a single surface or lens of non-zero power. Field curvature often goes by the name Petzval curvature.

The dependence of field curvature on lens characteristics is particularly interesting, in that for a thin lens of refractive index n and second focal length f', the radius of curvature r_p of the curved focal surface can be shown[38] to be $r_p = -nf'$. (The negative sign indicates that the focal surface of a positive lens curves to the left, i.e., the centre of curvature of the focal surface is to the left of the surface.) The sag of a spherical surface of radius r inspected not too far from the optical axis at a height h' is, as we have seen in other contexts, $\zeta =_p h'^2/(2r)$. For the Petzval surface, this gives a sag $\zeta_p =_p -h'^2/(2f'n)$, i.e., paraboloidal.

It was noted previously that the tangential astigmatic image surface was displaced from the unaberrated image surface by 3× the displacement of the sagittal astigmatic image surface. We can now identify the unaberrated image surface in these comparisons as the Petzval surface, so three paraboloids corresponding to the Petzval surface and the tangential and sagittal astigmatic surfaces are nested together; they coincide if astigmatism is zero, and separate into three different paraboloidal surfaces if astigmatism is present.

For a single thin lens in air, where the aperture stop coincides with the lens, the radii of curvature of the three focal surfaces are given relatively simply by[39,40] $r_s = -\dfrac{n}{n+1} f'$ for the sagittal surface, $r_t = -\dfrac{nf'}{3n+1}$ for the tangential surface, and (as above) $r_p = -nf'$ for the Petzval surface. The corresponding sag values are $\zeta_s = -\dfrac{n+1}{2n}\dfrac{h'^2}{f'}$, $\zeta_t = -\dfrac{3n+1}{2n}\dfrac{h'^2}{f'}$ and $\zeta_p = -\dfrac{1}{2n}\dfrac{h'^2}{f'}$, and the longitudinal astigmatism is therefore $\zeta_t - \zeta_s = -h'^2/f'$. Very obviously, this is non-zero (except on-axis where $h' = 0$), and does not depend on the location of the object or image, or the bending of the lens, as foreshadowed above, but fortuitously does depend on the sign of the lens power.

As a result of the similar impacts of astigmatism and field curvature on the shape of the focal surface, the Seidel aberration term for S_{III} (astigmatism) is usually grouped with that for S_{IV} (field curvature), and both have been tabulated together in Figure 3.7 (insets) for that reason. It is important to emphasise again, with an eye on image quality, that field curvature does not disrupt individual pencils forming the image, but astigmatism does by separating the tangential and sagittal foci, producing two elongated images orientated at 90 degrees to one another and blurred at axial locations in between.

We postpone further discussion of how to minimise astigmatism until we are more ready to discuss multi-lens systems. However, one glimmer of hope on the horizon might be appreciated now concerning field curvature. In a system comprising several lenses, the curvature (i.e. the reciprocal of the radius of curvature) of the Petzval surface is simply the sum of the Petzval curvature terms for the individual lenses:

$$\frac{-1}{r_P} = \frac{1}{n_1 f_1'} + \frac{1}{n_2 f_2'} + \ldots = \frac{F_1}{n_1} + \frac{F_2}{n_2} + \ldots \tag{3.8}$$

irrespective of where in the system each lens is situated. Noting that refractive index values vary only over a relatively small range, around 1.5–1.8, this sum can be thought of broadly as a sum of lens powers $\approx (F_1 + F_2 + \ldots)/1.6$. We recognise additionally that the use of higher refractive index glasses reduces the magnitude of the Petzval curvature term for a given focal power via the ratios F/n; this is another case where lanthanum crown glasses (Figure 2.3b) may be preferred, to achieve the required power without excessive monochromatic or chromatic aberrations.

Recall the discussion of relay lenses and field lenses (Section 3.4) which showed that a lens very close to one of the conjugates will not greatly affect the image formation. Nevertheless, such lenses will feature in the field curvature calculation above, and thus a judiciously placed negative-power lens can potentially function as a field-flattening lens, offsetting the Petzval curvature introduced by one or more positive-power lenses, without compromising the image-forming role of the system. This gives an optical designer scope to control the field curvature in an optical system by balancing positive and negative lenses almost independently of their role in image formation. We shall see in Chapter 5 that several modern eyepiece designs exploit this approach to reduce the field curvature that would otherwise result from a high-positive-powered eyepiece. Moreover, as astigmatism is predominantly dependent on lens power, the inclusion of a negative field-flattening lens also helps offset the astigmatism inherent in a positively-powered eyepiece.

Field Curvature and Visual Accommodation

While a flat field is essential for photography using a flat sensor, it is sometimes suggested that for visual observing, a flat field is less important because the eye has the ability to adjust focus for field curvature as it scans across the field of view. Field curvature is certainly less important than astigmatism, as the eye has no ability to reconcile the separated astigmatic tangential and sagittal images.[41] However, to make sense of the response of the human eye to field curvature, we need to understand a bit more about human vision.

The eye has two main light receptor cell types called rods and cones, named for their shapes. There are three cone types that are sensitive over different wavelength ranges: S (short wavelength) at the blue end, M (medium) in the mid-range and L (long wavelength) further towards the red end, peaking at 420 nm, 534 nm, and 563 nm, respectively.[42] Because the three cone types have different wavelength sensitivities, they provide information to the brain on the relative intensities of light at different wavelengths, and this is the basis of colour perception. The rod cells, on the other hand, are of a single type and thus provide no colour perception.

Rods and cones have very different distributions across the retina. Cone cells are strongly concentrated in a small portion of the retina called the fovea, just 1.5 mm across. Moreover, foveal cones are even slimmer than those further away; they are less than one-third the size in the centre of the fovea, 2.2 microns in diameter compared to 7.2 microns just 1.35 mm away, which allows for denser packing.[43,44] Rods, however, have a very low density in the fovea (zero in the centre) and predominantly populate the remainder of the retina where cones are less abundant. When the brain alerts us to images falling outside the fovea, the resolution is quite poor, partly due to the lower density of light-receptive cells, and partly because cells away from the fovea share nerve pathways to the brain, so the brain receives indistinct information about where the light is striking the retina and thus where exactly it is coming from. Resolution estimates around 1–2 arcmin for foveal vision and 5 arcmin outside the fovea are not uncommon. As a consequence, we only become aware of the details of an object when we shift the eyeball to bring that object's light onto the fovea. You're doing that right now, as you scan your eye along the line of text in this sentence to pick up the detailed patterns of successive letter shapes, probably reading only 5 to 10 letters, perhaps 1 or 2 syllables, at a time and relaying those to the brain to recognise and decode them.

Cones require a higher threshold of stimulation to register light, so they are effective when light is abundant, but they fail to register faint light which only the rods, principally away from the fovea, can detect.[45]

These various differences can be summarised thus: in bright-light conditions, the eye has excellent colour perception, which is concentrated in the finely sampled fovea, but in low-light conditions, the eye has no foveal sensitivity, which necessitates using averted vision, and no colour perception.

The eye has two main refractive elements: the front surface of the strongly curved cornea, and an internal crystalline lens that is rapidly reshaped under subconscious muscular control to adjust its focal power and hence to refocus the eye as required for different object distances. That refocussing activity is called accommodation,[46] but it is driven by cones, not rods, and is therefore ineffective under low light conditions.[47] So while your eye will be able to shift focus rapidly if you move your gaze from the bright disk of Jupiter, say, to one of its bright satellites in the periphery of the field of view, or from feature to feature across the bright image of the Moon, accommodation will not work at all if you are relying solely on rod-facilitated averted vision when observing a faint target. Moreover, in the absence of an accommodation stimulus from the cones, the eye does not relax to infinity but takes up a slightly myopic focal configuration; this behaviour is termed "night myopia".[48]

One implication of night myopia for astronomical observers is that while bright planetary targets and close double stars can be observed in quite sharp, colourful detail at the fovea, faint targets like diffuse galaxies tend to lack detail and colour in the peripheral field of vision. This is certainly exacerbated by the rods populating the peripheral area of the retina at a lower density, and multiple rod cells sharing nerve fibres away from the fovea, so the eye cannot tell precisely where the light it is sensing actually encountered the retina.[49] Accommodation of the eye on bright targets directed to the fovea can easily and quickly adjust for slight focussing errors of the telescope, even in the presence of a curved field. However, the eye cannot accommodate on faint objects, and if you thought you could instead focus on bright objects and then switch to faint ones, the eye will undermine your plans by going slightly myopic in the darkness! Actually, the story gets worse, as the ability to accommodate at all decreases steadily with age,[50] typically coming to an end (termed presbyopia) around age

50,[51,52] usually recognised when a person finds their arms have become too short to hold a book far enough away to focus on it, and they have to capitulate to the need for reading glasses. So, while accommodation of the eye may help reduce the impact of field curvature under bright visual conditions, this may be inadequate under low-light, astronomical conditions or if you are older than ~50 years of age.

The ability to accommodate as one's gaze shifts across the field of view has implications for the management of field curvature and astigmatism in an eyepiece. Assuming the observer focusses the telescope for infinity adjustment (or at least for their own far point[53]) using the best image formed by paraxial rays in the centre of the field, then as they scan their gaze around the curved field they will be able to accommodate to focus on images associated with the foreground, the pencils of which will be diverging when they reach the eye and can benefit from increased power provided by the eye's crystalline lens, but they will not be able to focus on images associated with the background, which produce converging pencils at the eye. If astigmatism cannot be *eliminated* in an eyepiece design, then in the *management* of astigmatism it is often therefore better to design the eyepiece so the sagittal surface is flat, i.e., in the same plane as the paraxial focus, while the tangential surface curves away from the eye.[54] The comparatively flat *telescope* focal plane will therefore place sharp images more or less in the paraxial plane, closer to the eye than the eyepiece's tangential focal surface. The tangential rays originating from the flat telescope focal plane will therefore arise closer to the eye than the eyepiece's tangential focal surface, and hence reach the eye diverging slightly, and the eye will be able to accommodate to focus them, or at least to focus on the smallest round image between the two – though either option necessitates defocussing the associated sagittal rays. This arrangement allows the eye to focus alternately on the sagittal or tangential rays or the point halfway between. If instead the tangential focal surface of the eyepiece had been flattened in the design, the sagittal surface would lie in the foreground, meaning the flat focal plane of the telescope would be behind the eyepiece's sagittal focal plane, and the sagittal rays would be converging when they reached the eye, which accommodation cannot refocus.

Distortion

The final Seidel wavefront term to consider in Equation 3.3 is $A_3 h'^3(4y_A)$, which gives rise to S_V. It is linear in meridional ray centration y_A, which corresponds to a tilt of the wavefront. The associated transverse aberration is given by application of Equation 3.4 which, including the *change* of sag ζ as a function of y_A as $\delta\zeta/\delta y$, makes the lateral aberration independent of y_A but still dependent on h'^3. So, like field curvature, distortion is the same for all rays in a given pencil and does not disrupt that pencil, but does move it to a different location in the focal surface. This transverse aberration can therefore be considered a change of lateral magnification with field angle, ever more so at higher field angles since it is proportional to h'^3. This results in image points having a different spatial relationship than the corresponding object points, with the image appearing stretched or compressed at greater field angles, hence the name distortion.

The human visual system (eye and brain combination) appears to be adept at re-interpreting images suffering from distortion.[55] Whereas most Seidel aberrations are quantified in mm or micron units, distortion is typically quoted as a percentage relative to the undistorted image. In many applications, the human eye can tolerate distortion of a few per cent; on paper, 4% distortion of a square is perceptible, and 10% becomes objectionable, so in general optical design, a threshold of up to 1% is regarded as reasonable.[56] In the case of visual observing of astronomical sources, the tolerance is probably greater[57] since many fields of view do not contain regular structures, so the brain will be less aware of distortion, and astronomical fields typically contain many areas devoid of structure (e.g. blank sky), so the eye–brain combination will be less affronted by a distorted image.

A single thin lens that is also the aperture stop is free of distortion, so the single lenses presented in Figures 3.6 and 3.7 show no evidence of this aberration (<0.1%), but it can become significant if the lens and aperture stop are separated, as is inevitably the case for eyepieces installed on a telescope, and therefore it must be monitored. (The aperture stop for an eyepiece in use on a telescope is the exit pupil of the main telescope optics, typically located some hundreds of millimetres upstream of the eyepiece.)

Distortion also has a different significance for an eyepiece than for, say, a photographic lens. The aim of a distortion-free photographic lens is to render a flat image of a field of objects, where a length in millimetres on the image can be associated with a scaled transverse length in the object space, and where the shapes of objects (such as rectangular buildings) are preserved. However, an eyepiece has to transform a more-or-less flat, uniformly scaled image in the focal plane of the telescope (with a scale of 1.15 arcmin per mm, for example), into a uniform *angular* distribution of principal rays when viewed through an eyepiece. This is all the more difficult because the focal length of the eyepiece is usually comparable to or even smaller than the diameter of the focal plane, as set by the field stop of the eyepiece. If you consider rays from various sources in the focal plane passing undeviated through the centre of a single, thin-lens eyepiece, perspective alone dictates that two points 1 mm apart near the centre of the focal plane will have a larger apparent angular separation to the observer than two points having the same 1 mm separation at the edge of the field of view. However, such an appearance would be unwelcome, as it would mean objects uniformly distributed in the focal plane (and thus uniformly distributed on the sky) would appear bunched up towards the edge of the field of view from the perspective of the observer. So, for visual observation through an eyepiece, the desire is *not* to have a distortion-free *linear-to-linear* spatial correspondence, but rather a distortion-free *linear-to-angular* transformation, i.e., the apparent angular separation of rays must increase towards the edge of the field of view to transform the uniform spatial separation of points in the focal plane to a uniform angular distribution of points as viewed through the eyepiece.

Aperture Stop Location

The aberrations a lens exhibits can, and often do, depend on the position of the aperture stop. Moving the aperture stop along the optical axis does not alter the refractive paths of the rays, so if a lens had no aberrations to start with, moving the aperture stop would not introduce any. However, moving the aperture stop redefines which ray is the principal ray, and which rays are included in the pencil. In the light of preceding discussions in this section, it should be clear that the appearance of many aberrations depends on the path of the principal ray and its location in the image plane, so redefining which ray is the principal ray changes which rays are admitted by the aperture stop and this makes a difference to the aberrations.

The order in which the Seidel aberrations were introduced in this section is material in this respect: a lens with residual spherical aberration will see a change in coma and subsequent aberrations if the aperture stop is moved, but if there is no spherical aberration, then moving the stop will not introduce coma. Likewise, if a lens has coma but no spherical aberration or astigmatism, astigmatism may be introduced from coma by moving the stop, but not spherical aberration. Such matters, called stop-shift effects, will not be discussed further here, and can be pursued elsewhere[58,59,60,61] if desired. Bypassing the details is not to suggest they are irrelevant, though, as a key feature of eyepiece design is that the aperture stop *is* separate from the lenses, as the entrance pupil of the eyepiece is the exit pupil of the main telescope optics. In the case of eyepiece design, stop-shift effects are something the designer has to contend with rather than being something that can be modified.

3.8 CHROMATIC ABERRATION

An inevitable consequence of refractive index depending on wavelength (Section 2.2) is that the power and focal length of a refracting surface, thin lens or thick lens also depend on wavelength, via the equations $F_{surf} = (n' - n)/r$, $F_{thin} = F_1 + F_2$, and $F_{thick} = F_1 + F_2 - (d/n)F_1F_2$. As dispersion cannot be avoided, the only means of counteracting it is to use a second lens to push divergent rays back together again in the image, matching the power of the system as a whole at two (or more) wavelengths. Usually, the Fraunhofer "F" and "C" lines, lying towards but not at the extremes of human wavelength sensitivity, are used for this purpose (see Figure 2.3a), though other combinations could be used in principle.[62]

Perhaps the best-known means of achieving this is to construct a doublet, i.e., a composite of two individual lenses made of different glass having different dispersive indices and different powers. The total power of the pair will be close to the sum of the two individual powers (it would be exact for two *thin* lenses in contact), so a strong positive lens and a weak negative lens will construct an overall positive doublet, but using a more dispersive (lower V-number) glass for the weaker negative lens means the dispersive properties of the two lenses can be cancelled, at least at two wavelengths. (A composite lens which accomplishes this feat at three wavelengths is said to be apochromatic.) Often the facing surfaces of the pair of lenses in a doublet will be designed with matching radii of curvature so they can be placed in complete contact to make a cemented doublet, though this is not a requirement; a split doublet can also perform the task, and the freeing-up of an extra radius of curvature and lens spacing provides two more degrees of freedom to the designer. The outer surfaces of cemented, achromatic doublets intended for observing distant objects normally resemble the best-form single lens in shape, i.e., almost plano-convex in the convex-first orientation, though numerous possible designs exist.[63,64]

We should, at this point, distinguish between chromatic aberration which manifests as lateral chromatic aberration and that which manifests as longitudinal chromatic aberration. As the adjectives suggest, lateral chromatic aberration produces a sideways separation of colour in the image, while longitudinal chromatic aberration moves the focal point along the optical axis as a function of wavelength. Both imply a change in magnification with wavelength. Lateral chromatic aberration means red and blue image points lie adjacent to one another, i.e., at slightly different angles from the optical axis in the image plane, and hence the image is smeared sideways like a spectrum. For longitudinal chromatic aberration, where the axial distance to the image plane depends on wavelength, the red pencils may still be converging when blue light is sharply imaged; this results in an oversized, poorly focussed red rim to a well-focussed blue image, and vice versa where the red image is well focussed. The images are more symmetric in the case of longitudinal chromatic aberration but still bear coloured rims.

Achieving chromatic correction via a doublet or similar two-glass solution is not the only way of addressing chromatic aberration. Perhaps unexpectedly, if two thin lenses of the same glass type are separated by a distance equal to the average of their focal lengths, lateral chromatic aberration will be eliminated.[65] To demonstrate how this comes about, the fundamental paraxial equation for a pair of separated thin lenses (Equation 2.17) can be written separately for two wavelengths, say for the Fraunhofer "F" and "C" spectral lines (see Section 2.2) as $F_{E,F} = F_{1F} + F_{2F} - dF_{1F}F_{2F}$ and $F_{E,C} = F_{1C} + F_{2C} - dF_{1C}F_{2C}$ where we recognise explicitly that the thin lens powers F_{1F}, F_{1C}, F_{2F} and F_{2C} are dependent on wavelength. We then demand that the overall system power (and hence effective focal length) is the same at both wavelengths by requiring $F_{E,F} = F_{E,C}$. Mathematically, we can then solve for the particular value of d which allows that to be achieved, and after a page of algebra and a few reasonable approximations, we find that the requirement is satisfied if $d = (f_1' + f_2')/2$, where the focal lengths are specified at an intermediate reference wavelength such as the "D" spectral line or a nearby wavelength. This solution is employed in the Huygens eyepiece, one of the two-lens eyepieces we meet in Chapter 4.

It is worth recalling (Equation 2.6) that when a particular surface power F is required, the radius of curvature of the surface r depends on the difference in the refractive indices of the two optical media at the surface, $n' - n$. In the case of an air-glass surface, where the difference is $n_g - 1$, a larger value of n_g permits a larger radius of curvature, i.e., a flatter surface. We have seen in Sections 3.5–3.7 that in general, monochromatic aberrations are smaller for flatter surfaces, as the angles of incidence and refraction then do not become so extreme. This is an argument in favour of using high-refractive-index glasses when possible. However, as Figure 2.3b shows, using a flint rather than traditional crown glass incurs a different penalty, that of increased dispersion and potentially worse chromatic aberration, unless the flint is being chosen as a negative component spe-cifically to offset chromatic aberration elsewhere. In a positive element, however, the choices point towards a high-refractive-index, low dispersion glass, like a lanthanum crown (see Figure 2.3b) to permit weaker surface curvatures *and* less dispersion, thus benefitting both monochromatic and chromatic aberration contributions.

In the chapters that follow, chromatic aberration will be quantified using the chromatic equivalents of the Seidel monochromatic wavefront coefficients, C_I and C_{II}, being the longitudinal and lateral chromatic wave-front aberrations.

3.9 ZERNIKE ABERRATION ANALYSIS

The Seidel description of aberrations, which was introduced in Sections 3.5–3.7 and which we utilise in much of this book, is not the only way to describe aberrations. Another approach originates with F. Zernike,[66] which we briefly introduce in this section. We will not utilise Zernike's approach in this book, so you could skip forward to the next chapter with-out undermining your understanding of the subsequent material, but ignoring the Zernike approach completely would be an unfortunate omis-sion from the narrative, so a brief description is provided.

The basis of the Zernike analysis is a series of mathematical functions that are defined especially for a circular domain (region of application), and which are thus well-suited to the passage of a wavefront through a circular aperture stop and lenses. In the Seidel analysis, each point within an aperture stop was defined with the coordinates (x_A, y_A), where the x-axis and y-axis are core to the Cartesian coordinate system. We often com-bined the coordinates to describe the centration, ρ, via $\rho^2 = x_A^2 + y_A^2$. The Zernike approach adopts a different coordinate framework, defining each point in the aperture not by (x_A, y_A) but according to its centration ρ and an angular coordinate φ (an azimuthal angle or "bearing" if you prefer) about the centre of the aperture. A coordinate system such as this, described by radial and angular measures (ρ, φ), is called a polar coordinate system, in contrast to the more familiar Cartesian coordinate system (x, y). The cen-tration is normalised to the radius of the aperture, so ρ ranges from 0 to 1.

The Zernike functions – polynomials – are a series of mathemati-cal expressions using the polar coordinates (ρ, φ) that separate into the product of two terms, the first being a radial term that depends only on the centration ρ, and the second being an angular term that depends only on the angular coordinate φ. The Zernike polynomials are typically symbolised by Z, while for convenience, the radial (centration-dependent) and angular (φ-dependent) terms can be called R and Φ, so we can write $Z(\rho,\varphi) = R(\rho)\Phi(\varphi)$. While this looks simple enough, there is a whole series of possible mathematical forms for $R(\rho)$, and similarly for $\Phi(\varphi)$. When added together with appropriate scaling factors, the Zernike polynomials allow any possible form of the wavefront, or more commonly the *devia-tion* of the wavefront from a spherical form, to be described mathemati-cally. Some of those mathematical forms resemble the now familiar Seidel descriptions, and many others appear unique to the Zernike approach. So, it is helpful to look more closely at the mathematical forms that $R(\rho)$ and $\Phi(\varphi)$ can take.

The angular terms are relatively simple, as the Zernike polynomial Z, and hence also Φ, must vary smoothly around the circular aperture and return to its starting value when φ moves through 360°. The angular func-tion therefore comes in only two forms: $\sin(m\varphi)$ and $\cos(m\varphi)$, where the sine function gives rise to a variation that is odd[67] about the axis defining $\varphi = 0$ and the cosine function gives rise to a variation that is even about the axis defining $\varphi = 0$. The non-negative integer m (=0, 1, 2, ...) determines how many cycles of variation occur as φ sweeps out a full circle around the aperture (i.e. going from 0° to 360°).

The angular coordinate must be defined relative to a reference direction, such as $\varphi = 0$ along the x-axis and, moving anti-clockwise, $\varphi = 90°$ on the y-axis; see inset to Figure 3.8. Alternative definitions may be made, such as defining the angle to be 0° at the y-axis and, moving clockwise, reaching 90° on the x-axis.[68,69] Such a change does not alter the set of functions, but switches labels between the odd and even angular functions.

The radial terms $R(\rho)$ are more varied and include a constant form ($R = 1$), a linearly varying form ($R = \rho$, recalling that $0 \leq \rho \leq 1$), and higher-order functions of ρ (quadratic, cubic etc.) which give rise to ripple-like radial patterns. To distinguish these different forms, they are labelled with the non-negative integers n and m as R_n^m, where n indicates the highest order of ρ in the function ($n = 0$ implies no variation with centration, $n = 1$ implies a linear dependence, etc.), and m is the order of the angular coor-dinate variation introduced in the paragraph above. The radial order n is further restricted: $n \geq m$, and $n-m$ must be even. Thus, the first few radial terms, including the two given above, are typically written $R_0^0 = 1$, $R_1^1 = \rho$, $R_2^0 = 2\rho^2 - 1$, $R_2^2 = \rho^2$ etc.

When the radial and angular terms R_n^m and Φ are reunited as the prod-uct $Z = R\Phi$, the various Zernike polynomials must also be labelled with

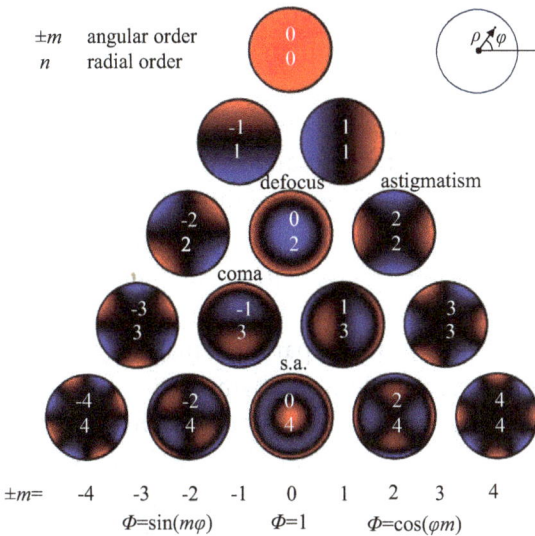

FIGURE 3.8 Visualisation of Zernike polynomials for radial orders $n = 0-4$. The first 15 Zernike polynomials are shown as colour plots (blue = −1, black = 0, red = +1)

n and m, as Z_n^m and Z_n^{-m} where the former, unsigned m values refer to the even, cosine forms of Φ and the latter, negative-signed m values refer to the odd, sine forms of Φ.

The Zernike polynomials can be easier to visualise in an illustration where the polynomials are calculated at each (ρ,φ) point within a circular aperture of radius $\rho = 1$, rather than as algebraic expressions. The polynomials for radial orders $n = 0, 1, 2, 3$ and 4 are illustrated in Figure 3.8.

Both the Zernike and Seidel approaches describe the aberrations of wavefronts, so we should expect some correspondence between the two methodologies, along with some differences. Recall that in the Seidel analysis, we defined the meridional plane as the yz-plane.

The simplest Zernike polynomial, Z_0^0, is uniform and corresponds to a uniform shift (sometimes called piston) of the wavefront along the optical axis. This case is almost trivial and does not appear in the Seidel analysis.

The two polynomials with $n = 1$, with a centration dependence ρ^1, have $m = 1$, giving a single cycle of angular variation and a linear radial variation, giving rise to functions which appear to tilt the wavefront about the x and y axes for the sine and cosine terms, respectively. Tilt will reposition but not defocus the pencil of rays associated with the wavefront, and if the tilt varies for pencils at different image heights as h'^3, then we will see the image points remain in focus, but the lateral magnification of the whole image will vary; this is what we called distortion in the Seidel analysis.

For $n = 2$, we see the Zernike wavefront variations that are commonly described as oblique astigmatism and defocus. We met ρ^2 terms in the Seidel analysis as oblique astigmatism and field (Petzval) curvature. The association between Zernike's "defocus" term and Petzval curvature is clear when the Zernike defocus is dependent on image height h'^2, for then the defocus changes progressively with field angle and curves the focal surface towards or away from the paraxial image plane.

At $n = 3$, we see a top-to-bottom asymmetry in the wavefront, which, with the ρ^3 dependence, represents Seidel's coma. Clearly, the Zernike analysis also points to other $n = 3$ wavefront aberrations besides. The richness of the Zernike polynomials is further evident at $n = 4$, where Z_4^0 is an axially symmetric term with a 4th-power dependence on centration ρ, which is what we called spherical aberration in the Seidel aberration, but there are additional ρ^4 polynomials outside the Seidel vocabulary.

The Zernick analysis obviously offers optical investigators a very powerful method of analysing the modulation of a wavefront as it passes through an optical system, but for an observer who is usually more interested in the resolution and clarity of an optical image presented by an eyepiece, it is sufficient to proceed with an understanding of Seidel aberrations.

NOTES

1 F.A. Jenkins and H.E. White, *Fundamentals of Optics*, 4th edition, McGraw-Hill, 1976, Chapter 21.

2 M.J. Kidger, *Fundamental Optical Design*, SPIE, 2000, Chapter 5.

3 It is interesting to note that accommodation is driven by the colour-sensitive cones of the eye rather than the rods, and thus for astronomers observing very faint fields with little activation of the cones, the eye settles to a more-or-less fixed focus distance corresponding not to infinity but to mild (~1 D) myopia.

4 W.N. Charman, in M.H. Freeman and C.C. Hull, *Optics*, 11th edition, 2003, Butterworth Heinemann, Chapter 15.

5 R. Kingslake and R.B. Johnson, *Lens Design Fundamentals*, 2nd edition, 2010, Academic Press, Chapter 1.

6 R. Kingslake and R.B. Johnson, *Lens Design Fundamentals*, 2nd edition, 2010, Academic Press, Chapter 1.

7 R. Kingslake and R.B. Johnson, *Lens Design Fundamentals*, 2nd edition, 2010, Academic Press, Chapter 2.

8 Begin by taking a factor of r^2 out of the square-root sign. As the value of y cannot exceed the radius of curvature r of the surface, $y/r < 1$, and the square-root term can be expanded using the binomial expansion $(1 + x)^p = 1 + px + [(p)(p-1)/2!]x^2 + [(p)(p-1)(p-2)/3!]x^3 + \cdots$

9 M.H. Freeman and C.C. Hull, *Optics*, 11th edition, 2003, Butterworth Heinemann, Chapter 7.

10 M. Born and E. Wolf, *Principles of Optics*, 7th edition, Cambridge University Press (1999), Chapter 5.

11 M.H. Freeman and C.C. Hull, *Optics*, 11th edition, 2003, Butterworth Heinemann, Chapter 7.

12 M.H. Freeman and C.C. Hull, *Optics*, 11th edition, 2003, Butterworth Heinemann, Chapter 7.

13 F.L. Pedrotti, L.M. Pedrotti, and L.S. Pedrotti, *Introduction to Optics*, 3rd edition, Pearsons (2007), Chapter 20.

14 M. Born and E. Wolf, *Principles of Optics*, 7th edition, Cambridge University Press (1999), Chapter 5.

15 D.J. Schroeder, *Astronomical Optics*, 2nd edition, Academic Press, 2000, Chapter 5.

16 M.H. Freeman and C.C. Hull, *Optics*, 11th edition, 2003, Butterworth Heinemann, Chapter 9.

17 M. Born and E. Wolf, *Principles of Optics*, 7th edition, Cambridge University Press (1999), Chapter 5.

18 M.J. Kidger, *Fundamental Optical Design*, SPIE, 2000, Chapter 6.

19 D.A. Atchison and G. Smith, *Optics of the Human Eye*, 2nd edition, 2023, Appendix 2.

20 D.J. Schroeder, *Astronomical Optics*, 2nd edition, Academic Press, 2000, Chapter 5.

21 M.J. Kidger, *Fundamental Optical Design*, SPIE, 2000, Chapter 6.

22 F.L. Pedrotti, L.M. Pedrotti, and L.S. Pedrotti, *Introduction to Optics*, 3rd edition, Pearsons (2007), Chapter 20.

23 E. Hecht, *Optics*, 5th edition, 2017, Pearsons, Chapter 6, with notation adjusted to the Cartesian framework adopted herein.

24 The one exception to this is the specific symmetric configuration where the object and image distances for the lens are the same, which arises only in a particular case called a "4f relay".

25 F.A. Jenkins and H.E. White, *Fundamentals of Optics*, 4th edition, McGraw-Hill, 1976, Chapter 9.

26 F.L. Pedrotti, L.M. Pedrotti, and L.S. Pedrotti, *Introduction to Optics*, 3rd edition, Pearsons (2007), Chapter 20.

27 F.A. Jenkins and H.E. White, *Fundamentals of Optics*, 4th edition, McGraw-Hill, 1976, Chapter 9.

28 F.A. Jenkins and H.E. White, *Fundamentals of Optics*, 4th edition, McGraw-Hill, 1976, Chapter 9.

29 M. Born and E. Wolf, *Principles of Optics*, 7th edition, Cambridge University Press (1999), Chapter 5.

30 F.L. Pedrotti, L.M. Pedrotti, and L.S. Pedrotti, *Introduction to Optics*, 3rd edition, Pearsons (2007), Chapter 20.

31 F.A. Jenkins and H.E. White, *Fundamentals of Optics*, 4th edition, McGraw-Hill, 1976, Chapter 9.

32 M.J. Kidger, *Fundamental Optical Design*, SPIE, 2000, Chapter 7.

33 F.A. Jenkins and H.E. White, *Fundamentals of Optics*, 4th edition, McGraw-Hill, 1976, Chapter 9.

34 J. Braat, The Abbe sine conditions and related imaging conditions in geometrical optics, *Proc. SPIE*, 3190, 59, 1997 https://spie.org/etop/1997/59_1.pdf.

35 Seidel astigmatism, also known as oblique astigmatism since it affects off-axis rays arriving at the lens at some oblique angle, is different from the common human eye condition that is also called astigmatism. The latter is due to some human eyes having different powers on different axes, in which case a given eye may have two different focal lengths depending on the portions of the eye the light passes through. This form of astigmatism can be thought of as the eye having a cylindrical variation to its roughly spherical symmetry, and it therefore requires a cylindrical variation in the spectacle lenses that may be prescribed to improve the vision (e.g. D.A. Atchison and G. Smith, *Optics of the Human Eye*, 2nd edition, CRC Press, 2000, Chapter 7). A similar instance of astigmatism may result if an optical element is deformed slightly by being over-tightened ("pinched") in its mount/cell.

36 F.A. Jenkins and H.E. White, *Fundamentals of Optics*, 4th edition, McGraw-Hill, 1976, Chapter 9.

37 M.J. Kidger, *Fundamental Optical Design*, SPIE, 2000, Chapter 6.

38 M.H. Freeman and C.C. Hull, *Optics*, 11th edition, 2003, Butterworth Heinemann, Chapter 9.

39 M. Born and E. Wolf, *Principles of Optics*, 7th edition, Cambridge University Press (1999), Chapter 5.

40 Born and Wolf (Note 9) define the radii R_s, R_t and R of the sagittal, tangential and Petzval surfaces as positive if they curve to the left, as arises for a positive powered lens, whereas I adopt the same convention as for the radii of curvature of a lens surface which is that they are negative if the centre of curvature is to the left of the intercept with the optical axis, and thus my variables r_s, r_t and r_p correspond to $-R_s$, $-R_t$ and $-R$, respectively.

41 W.N. Charman, in M.H. Freeman and C.C. Hull, *Optics*, 11th edition, 2003, Butterworth Heinemann, Chapter 15.

42 M.H. Freeman and C.C. Hull, *Optics*, 11th edition, 2003, Butterworth Heinemann, Chapter 10.

43 See Endnote 3

44 D.A. Atchison and G. Smith, *Optics of the Human Eye*, 2nd edition, CRC Press, 2000, Chapter 1.

45 W.N. Charman, in M.H. Freeman and C.C. Hull, *Optics*, 11th edition, 2003, Butterworth Heinemann, Chapter 15.

46 D.A. Atchison and G. Smith, *Optics of the Human Eye*, 2nd edition, CRC Press, 2000, Chapter 2.

47 W.N. Charman, in M.H. Freeman and C.C. Hull, *Optics*, 11th edition, 2003, Butterworth Heinemann, Chapter 15.

48 R.B. Rabbetts, *Bennett and Rabbetts' Clinical Visual Optics*, 4th edition, Butterworth Heinemann, 2007, Chapter 7.

49 R.B. Rabbetts, *Bennett and Rabbetts' Clinical Visual Optics*, 4th edition, Butterworth Heinemann, 2007, Chapter 3.

50 D.A. Atchison and G. Smith, *Optics of the Human Eye*, 2nd edition, CRC Press, 2000, Chapter 1.

51 D.A. Atchison and G. Smith, *Optics of the Human Eye*, 2nd edition, CRC Press, 2000, Chapter 2.

52 D.A. Atchison and G. Smith, *Optics of the Human Eye*, 2nd edition, CRC Press, 2000, Chapter 20.

53 The far point of the eye is the point in the distance from which light rays will be focussed perfectly onto the retina by the unaccommodated eye. For someone with perfect vision, the far point would normally be at infinity, but for a person with myopic eyes, the far point is in the foreground. As accommodation allows someone to adjust the focus of their eyes for objects closer than their far point, but not beyond, a myopic individual cannot see distant objects clearly, and accommodation cannot help; a negative-power spectacle lens is required to achieve that, e.g., see M.H. Freeman and C.C. Hull, *Optics*, 11th edition, 2003, Butterworth Heinemann, Chapter 6; R.B. Rabbetts, *Bennett and Rabbetts' Clinical Visual Optics*, 4th edition, Butterworth Heinemann, 2007, Chapter 4; D.A. Atchison and G. Smith, *Optics of the Human Eye*, 2nd edition, CRC Press, 2000, Chapter 7.

54 D.A. Atchison and G. Smith, *Optics of the Human Eye*, 2nd edition, CRC Press, 2000, Chapter 16.

55 W.N. Charman, in M.H. Freeman and C.C. Hull, *Optics*, 11th edition, 2003, Butterworth Heinemann, Chapter 15.

56 R. Kingslake and R.B. Johnson, *Lens Design Fundamentals*, 2nd edition, 2010, Academic Press, Chapter 11.

57 A. Nagler, Plössl Type Eyepiece for Use in Astronomical Instruments, US Patent 4482217, 1984, https://worldwide.espacenet.com/patent/search/family/023857522/publication/US4482217A?q=pn%3Dus4482217 (accessed 23/02/2025).

58 M. Born and E. Wolf, *Principles of Optics*, 7th edition, Cambridge University Press (1999), Chapter 5.

59 M.H. Freeman and C.C. Hull, *Optics*, 11th edition, 2003, Butterworth Heinemann, Chapter 7.

60 D.J. Schroeder, *Astronomical Optics*, 2nd edition, Academic Press, 2000, Chapter 5.

61 R. Kingslake and R.B. Johnson, *Lens Design Fundamentals*, 2nd edition, 2010, Academic Press, Chapter 12.

62 The "D" and "E" lines provide one such alternative pairing; T. Townsend Smith, The choice of wave-lengths for achromatism in telescopes, *Nature*, 114, 536 (1924) https://www.nature.com/articles/114536a0.

63 T. Townsend Smith, The color correction of an achromatic doublet, *Journal of the Optical Society of America*, 10, 39 (1925) https://opg.optica.org/josa/abstract.cfm?uri=josa-10-1-39.

64 M.H. Freeman and C.C. Hull, *Optics*, 11th edition, 2003, Butterworth Heinemann, Chapter 14.

65 T. Townsend Smith, The color correction of an achromatic doublet, *Journal of the Optical Society of America*, 10, 39 (1925) https://opg.optica.org/josa/abstract.cfm?uri=josa-10-1-39.

66 F. Zernike, Beugungstheorie des schneidenver-fahrens und seiner verbesserten form, der phasenkontrastmethode, *Physica*, 1, 689, 1934. https://www.sciencedirect.com/science/article/abs/pii/S0031891434802595 (accessed 12/04/2025).

67 "Odd" and "even" in this context refer to the mathematical definitions of a function f as odd if $f(-x) = -f(x)$ and even if $f(-x) = f(x)$.

68 M. Born and E. Wolf, *Principles of Optics*, 7th edition, Cambridge University Press (1999), Chapter 9.

69 F.L. Pedrotti, L.M. Pedrotti, and L.S. Pedrotti, *Introduction to Optics*, 3rd edition, Pearsons (2007), Chapter 20.

CHAPTER 4

Two-Lens Eyepieces

W<small>E FOUND IN</small> C<small>HAPTER</small> 3 that a single-lens eyepiece fails to deliver our design ambition of a 25 mm focal length optic capable of delivering images of stars to arcsecond resolution over a 32 arcmin field of view. The eyepiece failed on many fronts, struggling to acquire a large enough field of view because of the small radii of curvature that a single lens would need to achieve this focal length, and certainly failing to deliver the image quality due to the presence of aberrations. We saw how spherical aberration and coma could be controlled to some degree by bending the lens, and how we might potentially weaken the lens surfaces and reduce aberrations further if we shared the required power across two or more lenses. Introducing additional lenses could also give us an opportunity to address other aberrations as well, such as field curvature, chromatic aberration and astigmatism for which merely bending the lens has little impact. We also saw that by introducing a field lens, we might redirect the rays of off-axis objects back towards the optical axis, reducing the sizes of the lens(es) that would be required to deliver the desired field of view. It is now time to increase the number of lenses we use, to begin to test the reality of the potential improvements foreshadowed above.

4.1 INCORPORATING A FIELD LENS

We begin by considering the possibility of locating a relay lens (field lens) in the focal plane of the telescope, i.e., where the main optics of the telescope have produced a real image of the sky which we now seek to convey to the eye in a magnified form. Recall from Section 3.4 that a relay lens in the plane of an object or image (conjugate) does not alter the location

DOI: 10.1201/9781003670506-4

or magnification of the conjugate, nor does it alter the cone angle of each pencil of rays passing through it (which in our example system are $f/10$ pencils), but instead it changes the angle at which each pencil is directed relative to the optical axis. This concept may seem to jar with perceived common sense; surely if a field lens redirects the principal rays, then to attain the desired magnification (120×; see Section 3.1) we need to weaken the power of the second lens? Not so! Reviewing the effective focal length and power of a pair of separated thin lenses (Equations 2.14 and 2.17) shows that if we seek an effective focal length $f'_E = 25$ mm from a pair of lenses, then we need $F_E = 40$ D. We can achieve this with two lenses of powers F_1 (the field lens) and F_2 (the second lens), and separation d, such that $F_1 + F_2 - dF_1F_2 = 40$ D. As the telescope focal plane is at a distance $l = 0$ from the field lens, the field lens produces an image at $l' = 0$, i.e., unshifted (see the fundamental paraxial equation for a thin lens, Equation 2.7). For our two-lens eyepiece to operate in infinity adjustment and produce parallel rays in each outgoing pencil, the second lens must be positioned its own focal length from the focal plane, i.e., $d = f'_2$. However, f'_2 is just $1/F_2$, so the requirement above for lens powers becomes $F_1 + F_2 - (1/F_2)F_1F_2 = 40$ D, i.e., $F_1 + F_2 - F_1 = 40$ D, giving $F_2 = 40$ D, which is unchanged from Chapter 3 when we didn't use the field lens. Therefore, the use of a field lens hasn't changed the power requirement for the second lens, but it will allow it to have a smaller physical radius as it does not need to intercept such widely divergent pencils, and this means the rays pass through at smaller centrations where the aberrations should be less. The lens can also be reduced in thickness. The second lens in this configuration is referred to as the eye lens, due to its proximity to the eye of the observer.

The analysis above confirms the requirement for F_2 but gives us little insight into a sensible choice for F_1, so it is timely to consider the impact of the field lens on the eye relief (Section 3.1) of the telescope+eyepiece combination. Using a single-lens, symmetric eyepiece as in Chapter 3, the eye relief will be essentially the second-vertex focal length (Equation 2.15) of the eyepiece, which for a symmetrical thick lens of thickness $d \sim 15$ mm, refractive index $n \sim 1.5$, and focal length $f' \sim 25$ mm is around 19.4 mm. (For a *thin* lens of this focal length, it would of course be 25 mm.) When a positive field lens is introduced, it refracts the principal rays towards the optical axis and reduces the distance at which they cross the optical axis, i.e., the location of the exit pupil. The eye relief therefore depends on the choice of F_1. The second-vertex focal length of a system of two thin lenses with separation $d = 25$ mm and having $F_2 = 40$ D is given by $f'_V = 25$ mm $\times (1 - 0.025F_1)$.

For every positive choice for F_1 (the field lens power), the eye relief is therefore reduced relative to the single, thin-lens comparison, with stronger field lenses yielding shorter eye relief. Moreover, we have seen already that the single, thin-lens eye relief is optimistic compared to a single, thick-lens calculation, so we can expect the eye relief of a two-lens eyepiece to be reduced further with an exact analysis.

So, along with the potential benefits of installing a field lens, we have also come across one of the practical downsides: including a positive field lens reduces the eye relief, which may or may not end up being a tolerable loss. There is, in this slightly unwelcome result, a glimmer of hope: the analysis suggests – correctly – that a negative powered field lens ($F_1 < 0$), though counter to the intended purpose at *this* stage, would have the effect of *increasing* the eye relief to a value greater than the effective focal length. Achieving this would require greater lens dimensions to capture the light and redirect it to the exit pupil but nevertheless foreshadows that vastly different eyepiece designs may deliver other benefits. We will return to this possibility in Chapter 5.

The impact of incorporating a field lens in the eyepiece previously shown in Figure 3.2 is shown in Figure 4.1, where the optical arrangements are the same in every respect except for the addition of a 100 mm focal length (+10 D) field lens to the 25 mm focal length (+40 D) eye lens in Figure 4.1b. As explained, the inclusion of the field lens to refract the off-axis pencils towards the optical axis does not change the power requirement for the eye lens, but the rays then pass through the eye lens closer to the optical axis, where they experience less deviation. This reduces the aberrations considerably for the off-axis field points and also allows the eye lens to be reduced in diameter and bent, improving the aberrations further, but unhelpfully reduces the eye relief.

4.2 REVERSED RAY TRACING

Before investigating different lens designs to improve the eyepiece further, it is useful to change the way the ray traces are calculated so we can visualise the aberrations better. It was noted in Chapter 2 that light paths are reversible, so optical systems can be traced back-to-front with no loss of accuracy. An eyepiece in infinity adjustment is designed to take light from a series of point source virtual objects in the focal plane of the telescope and render these as parallel pencils of light passing through the exit pupil of the telescope+eyepiece combination. We can equally trace an eyepiece starting with a series of pencils containing parallel rays approaching the

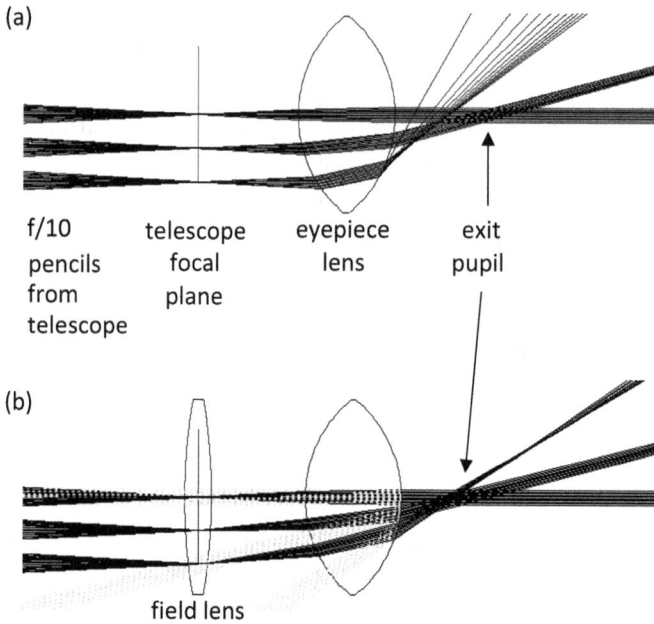

FIGURE 4.1 Ray paths for three object points through trial 25 mm focal length eyepieces. (a) As for Figure 3.2 (b) As for (a) but with the addition of a 100 mm focal length field lens in the focal plane of the telescope. The field lens does not alter the angular magnification of the configuration (see text), but the pencils for off-axis field points now pass through the eye lens at lower centration than in case (a). This greatly improves the imaging performance for off-axis field points as the deviation of the rays by the eyepiece is now shared by two lenses, even though the eye lens is unchanged. The eye lens could be reduced in radius and "bent" to reduce spherical aberration and reduced in thickness. One downside of the change is that the incorporation of the field lens in (b) decreases the eye relief (distance between the eye lens and telescope+eyepiece exit pupil).

eye lens from (negative) infinity, pass the rays through the eyepiece from eye lens to field lens and on to the telescope focal plane where they form focussed images, and then onwards to exit the telescope via its aperture stop (or at least the exit pupil of the telescope's main optics not including the eyepiece). The ray trace only shows those rays which successfully pass out through the defined aperture stop, so despite tracing it back-to-front, we see the same set of rays that would have passed from the telescope to the eyepiece in the conventional sense – almost, at least.

One *difference* between these two approaches is that even in infinity adjustment, not *all* rays in an incoming pencil exit the system parallel to the principal ray, because of aberrations. Consequently, modelling the system in reverse by directing a parallel pencil of rays into the eye lens will trace a slightly different set of rays than would have entered in the conventional sense, but then these rays won't intersect perfectly in the telescope focal plane, so we will still be able to assess the size and nature of the aberrations that arise in the eyepiece. The benefits of being able to inspect the ray aberrations for the eyepiece via calculations and spot diagrams make this reversed approach greatly preferred, and it is the one most optical designers have adopted in recent decades.

Another difference when tracing rays in reverse relates to the location of the aperture stop in the ray trace. Instead of using the exit pupil of the main telescope optics, which clearly differs from one telescope to another, a common approach in eyepiece design is to adopt a comparable telescope+eyepiece exit pupil as the entrance pupil of the reversed system,[1] typically sized at around 5 mm diameter since this corresponds to a fully dilated pupil of an older adult observer.[2] These two choices are equivalent in the paraxial regime, but not in the presence of aberrations. In particular, note that the telescope+eyepiece exit pupil is an *image* of the telescope entrance pupil (and of the exit pupil of the main optics of the telescope), formed by the eyepiece optics which are imperfect. Therefore, a real eyepiece forms the telescope+eyepiece exit pupil as an inevitably *aberrated* image of the telescope entrance pupil, and consequently, it is not an exact conjugate of the entrance pupil for non-paraxial rays. Tracing the eyepiece back-to-front with an idealised exit pupil tends to mask the aberration of the exit pupil, which is problematic since spherical aberration of the exit pupil can be severe (as we will see later). For this reason, I present ray traces and analyses for an aperture stop located at a dummy $f/10$ (and frequently also $f/6$) telescope exit pupil, rather than at the telescope+eyepiece exit pupil.[3] Inevitably, the choice of aperture-stop location affects the calculations of the Seidel aberrations and spot diagrams. This is obviously true in the many instances where I also show spot diagrams for $f/6$ pencils, and is likewise true for comparisons against other calculations by designers who adopt a different aperture stop size and location. While the details of outcomes will differ depending on the assumed stop size and location, the overall impression gained from the aberration calculations will still be valid for the $f/10$ or $f/6$ pencils as specified.

Reversed ray traces for the two lenses shown in Figure 4.1 are shown in Figure 4.2, along with spot diagrams for the two setups. Each small grid

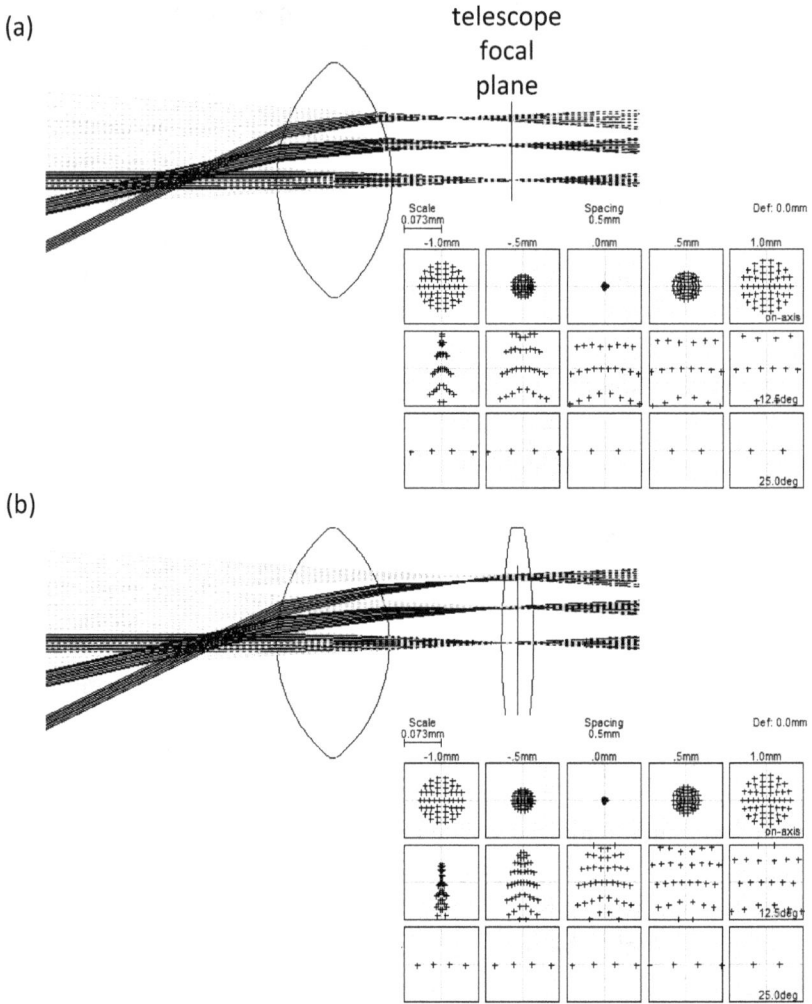

FIGURE 4.2 Reversed (left-to-right = eyepiece-to-telescope) ray traces for two eyepieces. The ray traces are for the same eyepiece designs as in Figure 4.1, for $f/10$ pencils associated with three apparent field angles 0°, 12.5° and 25°. Spot diagrams are shown in the inset for five defocus values. Each small grid in the spot diagrams corresponds to 1 arcsec for a 3000 mm effective focal length telescope. (a) 25 mm focal length symmetric lens. The on-axis image is excellent, as you would expect for essentially paraxial rays, but imaging of the off-axis field points is poor. (b) 25 mm effective focal length, using the eye lens as in (a) and a 100 mm focal length (10 D) symmetric field lens. The off-axis images are improved slightly, while the eye lens diameter and thickness can be reduced. The eye relief is smaller, though this could be partially increased by reducing the physical size of the eye lens.

square in the spot diagrams is 0.0145 mm wide, corresponding to 1 arcsec for a telescope with a 3000 mm effective focal length.

As noted previously, positioning the field lens in the focal plane is undesirable for the very practical reason that any dust or scratches on the surface of the lens will be in focus with the image and will detract from its appearance. We should therefore move the field lens slightly upstream (towards the main telescope optics) or downstream (towards the observer) to displace it from the focal plane. This offset lies at the heart of the first two "serious" eyepiece designs we review in the following two sections, respectively, the Huygens and Ramsden eyepieces.

Of course, shifting the field lens away from the focal plane immediately negates the condition discussed above that inserting a field lens in the focal plane does not require a modification of the power of the eye lens to deliver the desired effective focal length. As the distance of the field lens from the focal plane of the telescope will no longer be zero, i.e., $l_1 \neq 0$, the distance to the intermediate image produced by the field lens will also be non-zero: $l_1' \neq 0$. In order for the eyepiece to operate in infinity adjustment, the eye lens has to be placed its own focal length f_2 from the intermediate image produced by the field lens, but as $l_1' \neq 0$, this intermediate image no longer coincides with the field lens, and hence the separation of the two lenses will no longer be f_2. This means we have to redesign the eye lens to suit the displaced field lens, to recover the intended effective focal length.

4.3 HUYGENS EYEPIECE

The typical design constraints for a Huygens eyepiece are that the field lens and eye lens should be constructed of the same glass type, usually a common crown glass; that they should be plano-convex; and that the lenses should be separated by the average of their focal lengths in order that the eyepiece has no lateral chromatic aberration (Section 3.8). As the field lens is positioned upstream of the focal plane of the telescope, the near conjugate comes shortly after the field lens, and hence it is oriented in the convex-first orientation. The strength of the field lens is not strictly specified, though, as discussed in Section 4.2, its choice will affect the strength of the eye lens. The stronger the field lens, the more the principal rays are bent towards the optical axis, and the shorter the eye relief will be, so an overpowered field lens is to be avoided. Typical values adopted for Huygens eyepieces are field-lens focal lengths 1.5–3 times the eye-lens focal length.[4,5]

Writing these two constraints thus: $d = \left(f'_{FL} + f'_{EL} \right) / 2$ and $R \equiv F_{EL} / F_{FL}$, where R has been defined as the ratio of the power of the eye lens (F_{EL}) to that of the field lens (F_{FL}), we can use the equation for the equivalent power of a pair of separated thin lenses (Equation 2.17) $F_E = F_{FL} + F_{EL} - d F_{FL} F_{EL}$ to estimate the lens requirements. Half a page of algebra allows us to show that $F_{EL} = \dfrac{2R}{R+1} F_E$, which for values of R in the range 1.5–3 shows the eye lens must be 1.2–1.5 times stronger than the eyepiece as a whole. So, to design a 25 mm focal length Huygens eyepiece ($F_E = 40$ D), we would need a 48–60 D eye lens, paired with a 32 or 20 D field lens, respectively. It should be apparent that these are not going to resemble thin lenses.

Going further, we can estimate the eye relief as the second-vertex focal length of the eyepiece, which can be shown to be $f'_V = \dfrac{R-1}{2R} f'_E$. For $R = 1.5$–3, this implies very short eye-reliefs of only 1/6–1/3 of the eyepiece focal length, so potentially just a few millimetres. To reiterate, these algebraic estimates utilise the separated thin lens equation, and these lenses are certainly not thin lenses, but the calculations nevertheless foreshadow the difficulty of achieving a comfortable eye relief with Huygens eyepieces. The situation is better for Huygens eyepieces used on microscopes with finite-conjugate objectives, as the entrance pupil is then relatively nearby (typically around 200 mm away) and the exit pupil is formed a short additional distance behind the second-vertex focal point, but for an astronomical telescope with much longer focal lengths and the telescope exit pupil further away, the offsetting of the telescope+eyepiece exit pupil is minimal.

We can use a ray trace to examine the performance of a 25 mm focal length ($F_E = 40$ D) Huygens eyepiece, adopting $R = 2.5$. The equations above imply $F_{EL} = 1.43 F_E = 57.1$ D, $F_{FL} = 22.9$ D, and their equivalent thin-lens separation will be $d = 30.6$ mm. Adopting a plano-convex form for the field lens in the convex-first orientation, and adopting N-BK7 glass ($n = 1.5168$), gives the radius of curvature of its first surface as $r_{FL1} = 22.55$ mm. The lens diameter depends on the field it is intended to cover. Previous experience indicates that Huygens eyepieces deliver apparent[6] fields of view around 40°, i.e., ±20° from the optical axis. We inspect the performance out to 24°, which requires a lens physical radius of 12.5 mm (10.5 mm clear radius plus 2 mm additional as adopted in Section 3.1), which in turn implies a sag $\zeta_{FL1} = 3.8$ mm, to which we add an edge thickness of 1 mm giving $t_{FL} = 4.8$ mm. Attempting the same calculation for the eye lens, which is 2.5× stronger, runs into a problem: the inferred radius of curvature of its convex surface is r_{EL1} = 9.05 mm, which is impossible to deliver over a 12.5 mm lens radius. Fortunately, as the field lens refracts the incoming pencils towards the optical axis, the eye lens can be much smaller, perhaps half the diameter or less, which reduces the sag substantially. A trial radius of 7 mm implies a sag ζ_{EL1} = 3.3 mm, so we can try a 4.3 mm thick eye lens with an inter-lens spacing of 30.6 mm minus the average lens thickness, i.e., 26.0 mm. The eye lens is also used in a convex-first orientation; although the telescope focal plane lying in front of the eye lens might be regarded as a nearby conjugate, the exit pupil of the telescope+eyepiece lies just a few millimetres beyond the eyepiece and this too is a nearby conjugate, though a conjugate of the aperture stop rather than the sky. Ray tracing with the eye lens reversed rapidly confirms that the convex-first orientation is much better.

The ray trace and spot diagrams for the Huygens eyepiece described above are given in Figure 4.3, with the inter-lens spacing tweaked to 27.4 mm to obtain the intended focal length of 25 mm. For f/10 pencils, the image performance is good for the on-axis image, with the rays falling within ~0.5 arcsec of the centre, but for an 18° apparent field angle the Seidel summary for the ray tracing reports 0.05 mm of tangential coma, 1.7 mm of Petzval curvature (due to the two strongly convex surfaces), 0.1 mm separation of the tangential and sagittal astigmatic surfaces, and 4% distortion. The Petzval curvature is evident from both the ray tracing diagram and the monochromatic spot diagrams (Figure 4.3a and c) as the position of best focus moves progressively along the optical axis (to the left) for fields further off-axis.

Ray tracing in three colours (blue 486.1 nm, green 586.7 nm and red 656.3 nm) shows that the intended correction of *lateral* chromatic aberration was successful, as illustrated by the coloured images overlapping well, though significant *longitudinal* chromatic aberration persists, which results in symmetric coloured rims on all images.

Finally, the eye relief is only 5 mm, which is certainly below the comfortable range.

While the eyepiece performs quite well on-axis, its performance clearly degrades rapidly at higher field angles. Designers will therefore generally place a limiting aperture called a field stop in the focal plane to cut off apparent field angles where the image quality is poor. In the case of a Huygens eyepiece, the field is usually restricted to around 20° either side of the optical axis, i.e., to 40° apparent field of view.

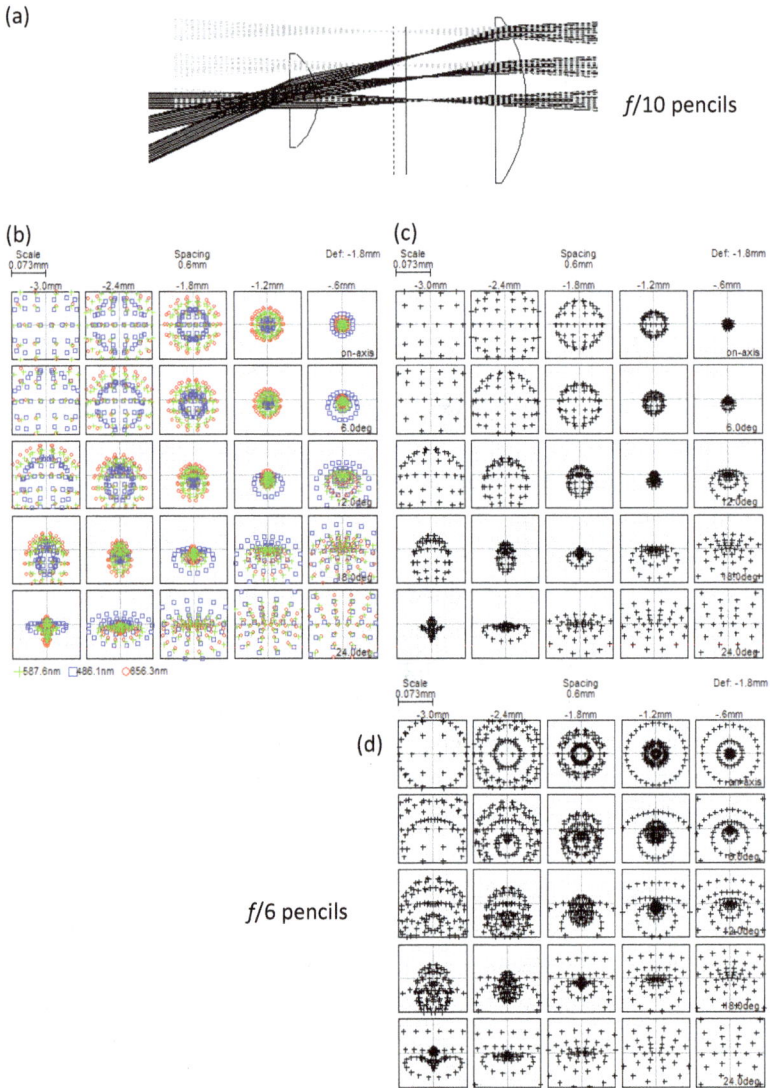

FIGURE 4.3   Huygens eyepiece (25 mm focal length, F_{FL} = 22.9 D, F_{EL} = 57.1 D). (a) Ray traces for apparent field angles of 0°, 12° and 24°. (b) Spot diagrams for three wavelengths at apparent field angles of 0°, 6°, 12°, 18° and 24°. Each small square corresponds to 1 arcsec in a 3000 mm focal length telescope. (c) Spot diagrams for $f/10$ pencils at apparent field angles of 0°, 6°, 12°, 18° and 24°. (d) Spot diagrams for $f/6$ pencils at apparent field angles of 0°, 6°, 12°, 18° and 24°. The image performance is good on-axis but deteriorates off-axis with significant Petzval curvature, coma and astigmatism. Huygens eyepieces are usually restricted to apparent field angles <20°, i.e., apparent field of view <40°. Performance for $f/6$ pencils is poor due to significant spherical aberration.

Until now, we have only examined f/10 pencils, but telescopes operating at "faster" f/ratios, such as f/6, will have pencils with larger cone angles, and the rays will cover larger areas on the eyepiece lenses they encounter. This, in turn, will subject those pencils to more extreme aberrations. We see that our Huygens eyepiece exhibits considerable spherical aberration once f/6 beams are introduced (Figure 4.3d), demonstrating the unsuitability of this eyepiece design for "fast" telescope optics.

Before considering an alternative eyepiece design, it is of historical interest to note that Christiaan Huygens developed his eyepiece with lateral chromatic correction in the early 1660s, more than 40 years *prior* to Newton publishing his major study on light, Opticks, in 1704. It was therefore achieved at a time when the understanding of dispersion was in its infancy, and Huygens' success in developing a chromatically corrected eyepiece is credited with spurring Newton to understand the science that enabled this feat to be achieved.[7]

4.4 RAMSDEN EYEPIECE

Whereas the Huygens eyepiece (Section 4.3) has the field lens displaced from the focal plane towards the main telescope optics, in the Ramsden eyepiece[8] it is displaced towards the eye lens, and because the focal plane (nearby conjugate) is now upstream of the field lens, the field lens is flipped over to the plano-first orientation, so the convex surfaces of the field lens and eye lens face one another. The same glass is used in both lenses, typically a common crown, and both have similar focal lengths.[9,10]

Starting with a thin-lens model, the lens spacing d must be somewhat less than the second focal length of the field lens f_{FL}', otherwise the exit pupil of the telescope+field-lens combination will fall at or in front of the eye lens. Setting the separation to a fraction g (gap fraction) of the field-lens focal length, $d \equiv g f_{FL}' = g/F_{FL}$ where $0 < g < 1$, the equations for a separated thin-lens pair imply $F_E = F_{FL} + F_{EL} - (g/F_{FL})F_{FL}F_{EL}$, i.e., $F_E = F_{FL} + (1 - g)F_{EL}$. For a case with equal focal lengths, i.e., $f_{FL}' = f_{EL}'$, the eyepiece power becomes $F_E = (2 - g)F_{EL}$ and consequently $F_{EL} = F_E/(2 - g)$. This is very significant: as $0 < g < 1$, it requires $F_{EL} < F_E$, i.e., the eye lens will be weaker than the eyepiece as a whole. This is a considerable improvement on the Huygens eyepiece, for which the eye lens had to be stronger than the eyepiece overall, and as this was the stronger of the two lenses, it accounted for most of the aberrations. If we adopt $g = 2/3$, for example, then $1/(2 - g) = 3/4$, so our 40 D Ramsden eyepiece will require a 30 D eye lens, compared to 57.1 D in the 40 D Huygens eyepiece in Figure 4.3.

Although the Ramsden eyepiece sacrifices the control of lateral chromatic aberration that we had in the Huygens eyepiece, we should now see a significant improvement in the monochromatic aberrations. Moreover, for this thin-lens Ramsden eyepiece we should expect an eye relief given by $f'_V = 1/F'_V$ where $F'_V = F_E/(1 - dF_{FL})$, i.e., $f'_V = f'_E/3 = 8$ mm. This is also an improvement on our Huygens eyepiece which has $f'_V = 5$ mm.

We can test this design as follows: an N-BK7, 30 D plano-convex eye lens requires a radius of curvature $r = 17.22$ mm. Desiring a thin lens separation of $(2/3)f'_{EL} = 22.2$ mm, and recognising that the principal planes of a plano-convex lens lie close to the convex vertex, we can adopt a trial inter-lens spacing of 22.2 mm, which delivers the desired focal length of 25 mm. One result we shall pre-empt is that the apparent field of view over which a Ramsden eyepiece delivers tolerable image quality, ~30°, is smaller than that for a Huygens eyepiece (~40°), and certainly less than the 60° field we requested in Section 3.1. Consequently, we will restrict the field for which we trace rays to 36°, which reduces the field in the image plane to a radius of 10.4 mm. This in turn implies a 12.4 mm lens radius, a sag of 5.3 mm, and a thickness $t = 6.3$ mm.

As with the Huygens eye lens, the Ramsden eye lens can be much smaller than the field lens, which allows it to be reduced further in diameter and thickness.

Compared to the Huygens eyepiece, the Ramsden eyepiece (Figure 4.4a) has much improved monochromatic aberrations (Figure 4.4c) and increased eye relief, though the chromatic aberrations (Figure 4.4b) are worse (as expected) and the useable field of view is smaller. However, even at the 18° apparent field angle where the Ramsden quality overall is poor, the Seidel summary indicates <0.01 mm of transverse spherical aberration (cf. 0.04 mm for the Huygens at 18°), 0.01 mm of tangential coma (cf. 0.05 mm), 1.3 mm of Petzval curvature (cf. 1.7 mm), and 3% distortion. However, the Ramsden eyepiece's astigmatism is much worse, as is evident from the spot diagrams, giving 1.2 mm separation of the tangential and sagittal astigmatic surfaces (cf. 0.1 mm for the Huygens eyepiece). Ray tracing in three colours shows, as expected, that lateral chromatic aberration is now worse at 0.07 mm and off-axis images are noticeably skewed in colour, whereas lateral chromatic aberration was corrected in the Huygens design (0.01 mm). However, the longitudinal chromatic aberration of the Ramsden eyepiece (0.31 mm) is half that of the Huygens eyepiece (0.65 mm). The Ramsden eyepiece also performs tolerably well with faster $f/6$ pencils (Figure 4.4e), which the Huygens eyepiece did not.

(a)

f/10 pencils

(b)   (c)

(d) Seidel wavefront
     aberrations

(e)

f/6 pencils

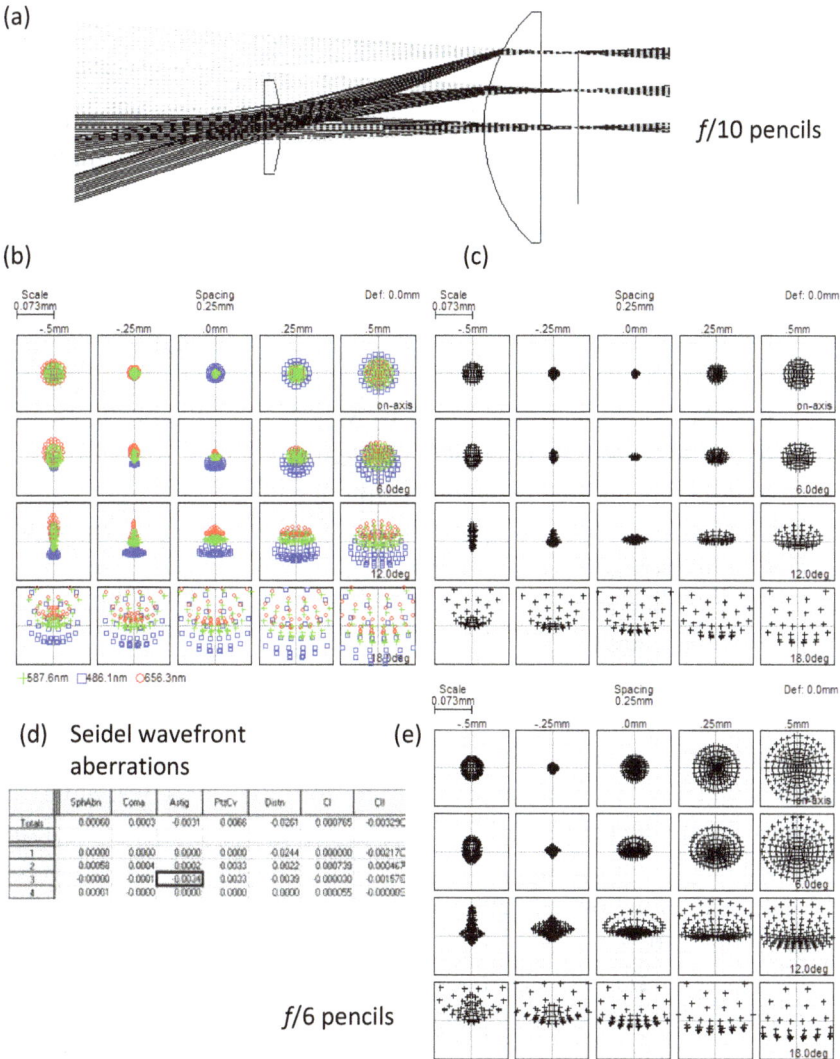

FIGURE 4.4 Ramsden eyepiece (25 mm focal length, F_{FL} = 30 D, F_{EL} = 30 D, d = 22.3 mm). (a) Ray traces for apparent field angles of 0°, 9° and 18°. (b) Spot diagrams for three wavelengths at apparent field angles of 0°, 6°, 12° and 18°. Each small square corresponds to 1 arcsec in a 3000 mm focal length telescope. (c) Spot diagrams for *f*/10 pencils at apparent field angles of 0°, 6°, 12° and 18°. (d) Seidel wavefront aberrations for each surface and the system as a whole. (e) Spot diagrams for *f*/6 pencils at apparent field angles of 0°, 6°, 12° and 18°. The image performance is good on-axis, and off-axis it experiences much reduced Petzval curvature compared to the Huygens eyepiece. Ramsden eyepieces are usually restricted to apparent field angles < 15°, i.e., apparent field of view <30°. Performance for *f*/6 pencils is better than for the Huygens eyepiece.

Overall, then, the Ramsden has some attractive features compared to the Huygens, but its main weaknesses are (1) strong lateral chromatic aberration and (2) a small field of view limited by the worsening of astigmatism at field angles exceeding around 15°. Whereas tangential coma increases linearly with field angle, astigmatism increases quadratically so limits the useable field more abruptly. In the quest for an eyepiece that delivers the desired properties (Section 3.1), these two aberrations are clearly of key importance to address. Inspection of the Seidel wavefront aberration calculations for each surface of the Ramsden eyepiece for the 18° field angle (Figure 4.4d) shows that the biggest contribution to astigmatism comes from the field lens (surface 3 of the back-to-front eyepiece layout, highlighted in Figure 4.4d). This might have been expected since this is the larger of the two lenses, and the 18° pencil passes through a steeply curved portion of this lens at high centration. Although the eye lens has the same power (30 D), the 18° pencil passes through a much flatter portion of this lens, closer to its centre. Therefore, it is the field lens that contributes most to the off-axis aberrations, so efforts to reduce astigmatism might sensibly focus on this surface first.

By relaxing the requirement that the two lenses in the Ramsden eyepiece have the same focal length, it is possible to improve the performance. Figure 4.5 shows a second 25 mm focal length Ramsden eyepiece design[11] (by M. Kidger) with the field lens weakened from 30 D to 24.3 D, while the eye lens is virtually unchanged at 29.3 D (formerly 30 D). To maintain the intended effective focal length (25 mm) and power (40 D), the separated thin-lens pair equation indicates that the separation d must be decreased from 22.2 to 19.0 mm.

The weakening of the plano-convex field lens clearly reduces the off-axis aberrations introduced there, evident both from the improvement in off-axis spot diagrams (Figure 4.5c) and the reduction of the Seidel wavefront astigmatism introduced at that surface to −0.0028 (formerly −0.0034; Figure 4.4d), reducing the distance between the tangential and sagittal astigmatic surfaces from 1.2 to 1.0 mm. The lateral chromatic aberration is made slightly worse, however.

Weakening the field lens also reduces the deflection of off-axis pencils at that lens, and this change, plus the reduction in the lens separation, means the pencils passing through the edge of the focal plane pass through the eye lens further from the optical axis. Consequently, the exit pupil of the telescope+eyepiece combination is formed further behind the eye lens, helpfully increasing the eye relief, in this case from 8 to 12 mm.

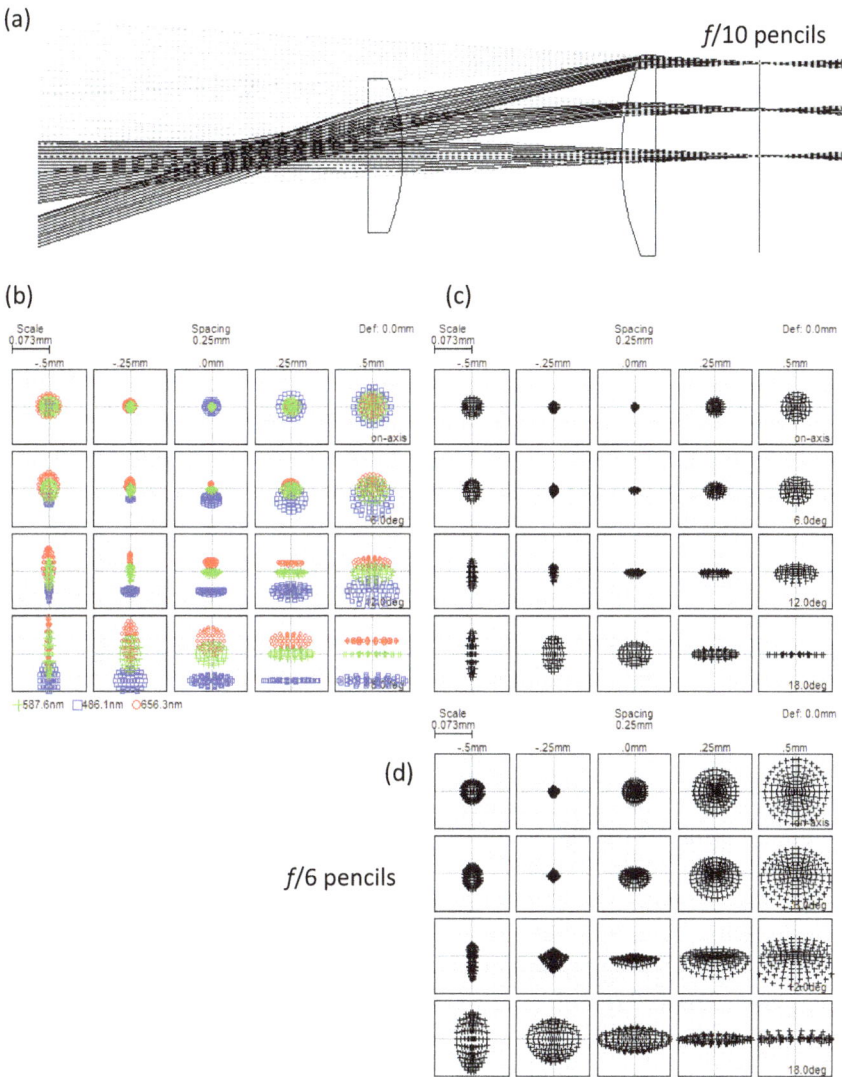

FIGURE 4.5 Ramsden eyepiece (25 mm focal length, F_{FL} = 24.3 D, F_{EL} = 29.3 D, d = 19 mm). (a) Ray traces for apparent field angles of 0°, 9° and 18°. (b) Spot diagrams for three wavelengths at apparent field angles of 0°, 6°, 12° and 18°. (c) Spot diagrams for f/10 pencils at apparent field angles of 0°, 6°, 12° and 18°. (d) Spot diagrams for f/6 pencils at apparent field angles of 0°, 6°, 12° and 18°.

The Ramsden eyepiece clearly offers better image quality than the Huygens eyepiece in some respects, but with poorer chromatic aberration and with astigmatism limiting the field of view considerably. Besides the Seidel wavefront terms, Figure 4.4d also shows the longitudinal chromatic (C_I) and lateral chromatic (C_{II}) wavefront aberration terms, and it is evident that these arise more at the eye lens (surfaces 1 and 2) than the field lens (surfaces 3 and 4). It is natural therefore to seek improvements in the chromatic performance by replacing the single eye lens with an achromatic doublet, pairing glasses of different dispersive index. This task is our next one, in Chapter 5, where we investigate the Kellner eyepiece and begin dealing with "modern" eyepieces utilising more than one glass type.

NOTES

1 M.J. Kidger, *Fundamental Optical Design*, 2000, SPIE, Chapter 11.

2 The dilated pupil diameter of the human eye under dark conditions depends on age, and is around 6–8 mm for 20 year olds, but decreases to around 4–5 mm for 80 year olds. A larger pupil also introduces larger aberrations, as we have seen for the Seidel aberrations which depend on ray centration. The aberrations of the eye noticeably limit image quality once the pupil exceeds around 2 mm. D.A. Atchison and G. Smith, *Optics of the Human Eye*, 2nd edition, 2023, CRC Press, Chapter 20. R.B. Rabbetts, *Clinical Visual Optics*, 4th edition, 2007, Butterworth Heinemann, Chapter 3.

3 For computational convenience in WinLens Basic, the *f*/10 configuration has been simulated using a 100 mm diameter aperture stop located 1000 mm from the telescope focal plane. Pencils for an *f*/6 system are simulated with a 166.7 mm diameter aperture stop in the same position.

4 F.A. Jenkins and H.E. White, *Fundamentals of Optics*, 4th edition, McGraw-Hill, 1976, Chapter 10.

5 M.H. Freeman and C.C. Hull, *Optics*, 11th edition, 2003, Butterworth Heinemann, Chapter 6.

6 "Apparent field angle" refers to the angle rays make to the optical axis in the image space where the visual observer places their eye. This terminology is consistent with the phrase "apparent field of view", which refers to the angular separation of opposite edges of the field of view as seen by the observer's eye. In contrast, the "true field of view" refers to the angle this corresponds to on the sky, which differs by a factor that is the angular magnification of the telescope+eyepiece combination (ignoring distortion, which changes the magnification as a function of field angle by typically a few percent).

7 R. Kingslake and R.B. Johnson, *Lens Design Fundamentals*, 2nd edition, 2010, SPIE, Chapter 5.

8 The Ramsden eyepiece was developed in 1782, 120 years after the Huygens eyepiece, by the mathematician and scientific instrument maker Jesse Ramsden, who was based in London at a time when the United Kingdom had a high demand for navigational instruments. He was married to the daughter of John Dolland who had successfully patented the achromatic doublet in 1758 (though certainly had not invented it). Despite this

association, it would be almost 70 years until Carl Kellner employed an achromatic doublet as the eye lens of the Ramsden eyepiece, producing the achromatic Ramsden or Kellner eyepiece in 1849; we consider this eyepiece in Chapter 5.

9 F.A. Jenkins and H.E. White, *Fundamentals of Optics*, 4th edition, McGraw-Hill, 1976, Chapter 10.

10 M.H. Freeman and C.C. Hull, *Optics*, 11th edition, 2003, Butterworth Heinemann, Chapter 6.

11 M.J. Kidger, *Fundamental Optical Design*, 2000, SPIE, Chapter 11.

Multi-Glass Eyepieces

5.1 ACHROMATIC RAMSDEN/KELLNER EYEPIECE

The Huygens and Ramsden eyepieces discussed in Chapter 4 benefitted from the simplicity of a single glass type, but suffered chromatic aberration (longitudinal for the Huygens, lateral for the Ramsden) along with astigmatism and other monochromatic aberrations. With chromatic aberration in the Ramsden eyepiece arising principally in the eye lens (Figure 4.5), it seems a logical step to explore whether replacing the eye lens with an achromatic doublet would result in a significant improvement. Doing so produces a related eyepiece known either as an Achromatic Ramsden or a Kellner eyepiece. The eyepiece was developed almost 70 years after the Ramsden by Carl Kellner, despite J. Ramsden presumably being well aware of the properties of achromatic doublets (see Chapter 4, Note 8), so I will refer to it as a Kellner.

Achromatisation

An achromatic doublet used as the objective lens in a refracting telescope is intended to take parallel light from a distant object and feed it to a focal plane free of chromatic aberration at two wavelengths. The design utilises a stronger positive crown lens and a weaker negative flint lens, where the stronger crown has the lower dispersion (high V-value), and the weaker flint has the higher dispersion (low V-value). For two *thin* lenses in contact, the total power is simply the sum of the two lens powers (Equation 2.17 with $d = 0$). For two *real* lenses in a cemented achromatic doublet, a very similar result holds, so using a stronger, positive crown lens and a weaker, negative flint lens results in a lens that is positive overall, while the higher

 DOI: 10.1201/9781003670506-5

dispersion of the weaker negative lens counteracts the lower dispersion of the stronger positive lens. Furthermore, because there are three optical surfaces to manipulate in a cemented doublet, the design can also be specified to minimise spherical aberration and coma. Four paraxial solutions exist[1] for achromatic crown+flint doublets (two crown-first, two flint-first), but the preferred form for aberration control when used to observe a very distant object is crown-first, of biconvex form. The accompanying negative flint is almost plano-concave (concave-first), so the doublet overall has a similar profile to a plano-convex lens used convex-first for a distant object to minimise spherical aberration and coma (Sections 3.6 and 3.7). In attempting to achromatise a Ramsden eye lens, the natural form to start with is perhaps a conventional achromatic doublet, and experience shows that plano-convex or almost plano-convex forms for the Kellner eye lens are indeed usually well suited.

We investigate three Kellner designs. Patents typically specify the glass types by refractive index n and Abbe number V rather than by glass name. In implementing the ray traces presented in this book, glasses having very similar n and V values have been adopted from the Qioptiq Glass Catalogue as implemented in the WinLens Basic ray-tracing programme, many of which are Schott glasses. It is recognised that the adopted glasses may not always match the specific types that the original designers had in mind, but they match adequately for our purposes.

The Kellner design in Figure 5.1 was detailed[2] in 1907 and is typical of the type. Both the field lens and the eye lens have outward facing plano surfaces. The chromatic performance of the eyepiece is certainly improved compared to the two Ramsden variants shown in Figures 4.4 and 4.5, to the extent that field angles out to 24° are shown for the Kellner (cf. 18° for the Ramsdens), but the Kellner images are still limited by astigmatism, as were the Ramsdens. As noted in Section 3.7, astigmatism in a thin lens (at the aperture stop) is proportional to the power of the lens but not its shape, so can't easily be addressed in a system with solely positive lenses.

It is evident from Figure 5.1a that the pencil of rays for the 24° field crosses the optical axis closer to the eye lens than the 12° pencil. In other words, the exit pupil suffers from spherical aberration. This pupil spherical aberration is formed in a similar way to spherical aberration of the images of the sky, but is distinct in that it affects the image of the aperture stop, which ideally should coincide with the pupil of the observer's eye. As with the focal points of a lens, the exit pupil is defined by the paraxial rays, so we can use the intersection of the principal 12° ray with the optical axis

(a)

$f/10$ pencils

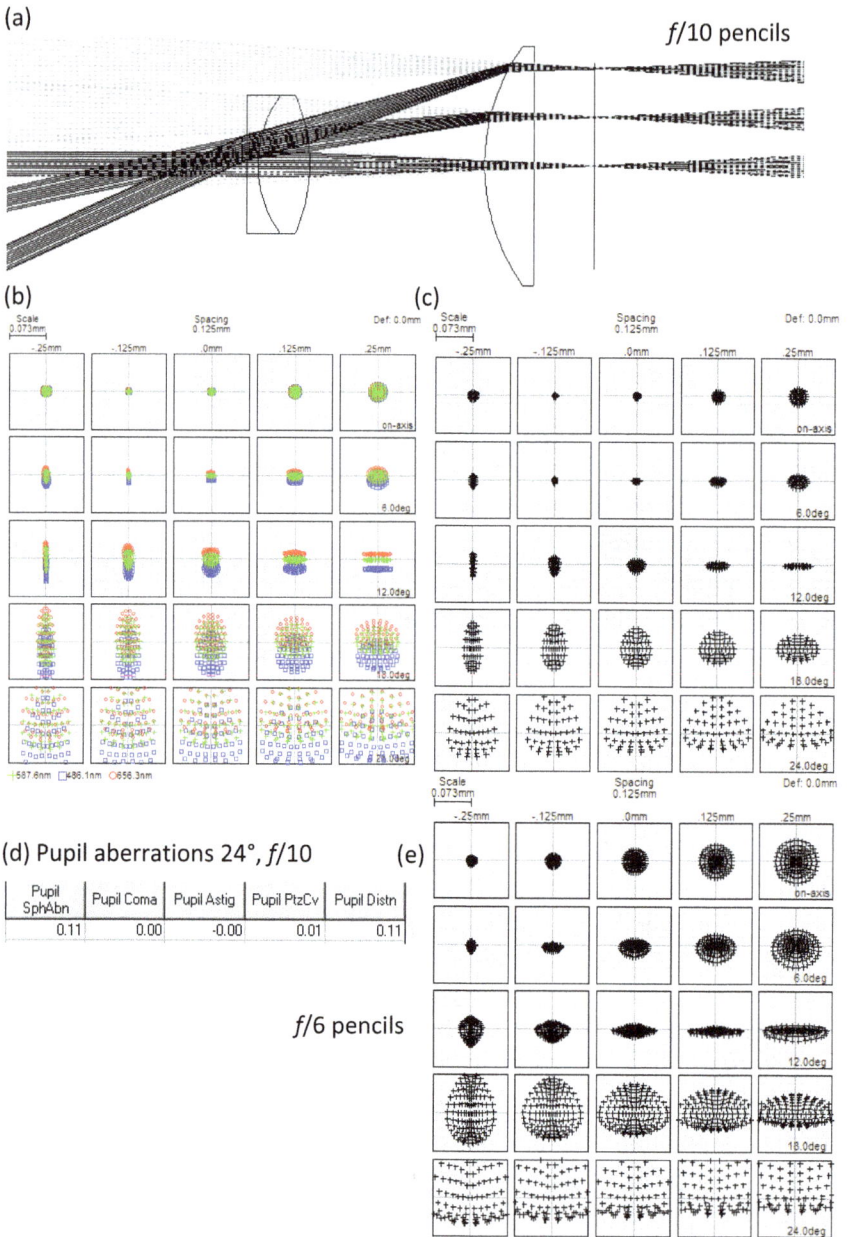

(b)

(c)

(d) Pupil aberrations 24°, $f/10$

(e)

Pupil SphAbn	Pupil Coma	Pupil Astig	Pupil PtzCv	Pupil Distn
0.11	0.00	-0.00	0.01	0.11

$f/6$ pencils

FIGURE 5.1 Kellner eyepiece for Carl Zeiss 1907 [25 mm focal length, F_{FL} = 23.2 D, F_{EL} = 30.4 D, d = 19.6 mm]. (a) Ray traces for apparent field angles of 0°, 12° and 24°. (b) Spot diagrams for three wavelengths at apparent field angles of 0°, 6°, 12°, 18° and 24°. (c) Spot diagrams for $f/10$ pencils at apparent field angles of 0°, 6°, 12°, 18° and 24°. (d) Pupil aberrations for eyepiece (24° field at $f/10$). (e) Spot diagrams for $f/6$ pencils at apparent field angles of 0°, 6°, 12°, 18° and 24°.

as a better indication of where the telescope+eyepiece exit pupil lies. What pupil spherical aberration means in practice is that an observer would have to move their head forward, closer to the eye lens, to intercept the 24° rays, and in so doing may start to lose sight, literally, of the rays at intermediate (e.g. 12°) field angles. This creates a patchy field of vision, sometimes referred to as a "kidney bean effect" based on the shape of the dark intermediate field angles in the visible image. Pupil spherical aberration[3] is tabulated in Figure 5.1d.

Figure 5.2 shows a second 25 mm focal length Kellner eyepiece of Albert König's 1953 design.[4] The cemented surfaces have the same radii of curvature as the convex surface of the field lens, assisting manufacture. An eye relief of 14 mm is achieved, and the pupil aberrations (Figure 5.2e) are much reduced compared to the 1907 design (Figure 5.1d). With a little defocus, the eyepiece would also perform adequately on an $f/6$ telescope, though with a little loss of the useable field of view due to a deterioration in the off-axis aberrations.

Michael Kidger presents an alternative design[5] which allows the external surface of the cemented doublet to take a weak convex form and utilises a relatively high refractive index glass for the crown (SK16, $n = 1.62$, $V = 60$) but lower than SF10 for the flint (SF5, $n = 1.67$, $V = 32$). The difference in refractive index between the two glasses is therefore less than in König's Kellner. An eye relief of 10 mm is obtained (Figure 5.3).

These three interpretations of Kellner eyepieces by expert optical designers of the 20th century begin to indicate the range of configurations that may be adopted to generate an eyepiece of a particular focal length (in this case, 25 mm) and type (in this case, Keller), but with different implementations and image characteristics. I would venture to suggest that König achieves better monochromatic correction, but Kidger achieves better lateral colour. Both eyepieces are limited in off-axis performance by astigmatism at field angles exceeding about 15°, and Kidger notes the existence of higher order coma at greater angles in his design.

With astigmatism prevalent in all three Kellner designs, it is natural to consider whether other design choices could have been made to help counteract this. Figure 5.2d shows that König's Kellner has −0.0054 mm of astigmatism (S_{III}), with most of that coming from the cemented doublet eye lens. Replacing the plano-convex *field* lens with a symmetric biconvex lens (N-BK7, $r_1 = 44.70$ mm, $r_2 = -r_1$, $d = 5.8$ mm) retains the 25 mm focal length and 14 mm eye relief but diminishes the astigmatism ($S_{III} = -0.0033$) without significantly changing other aberrations. The

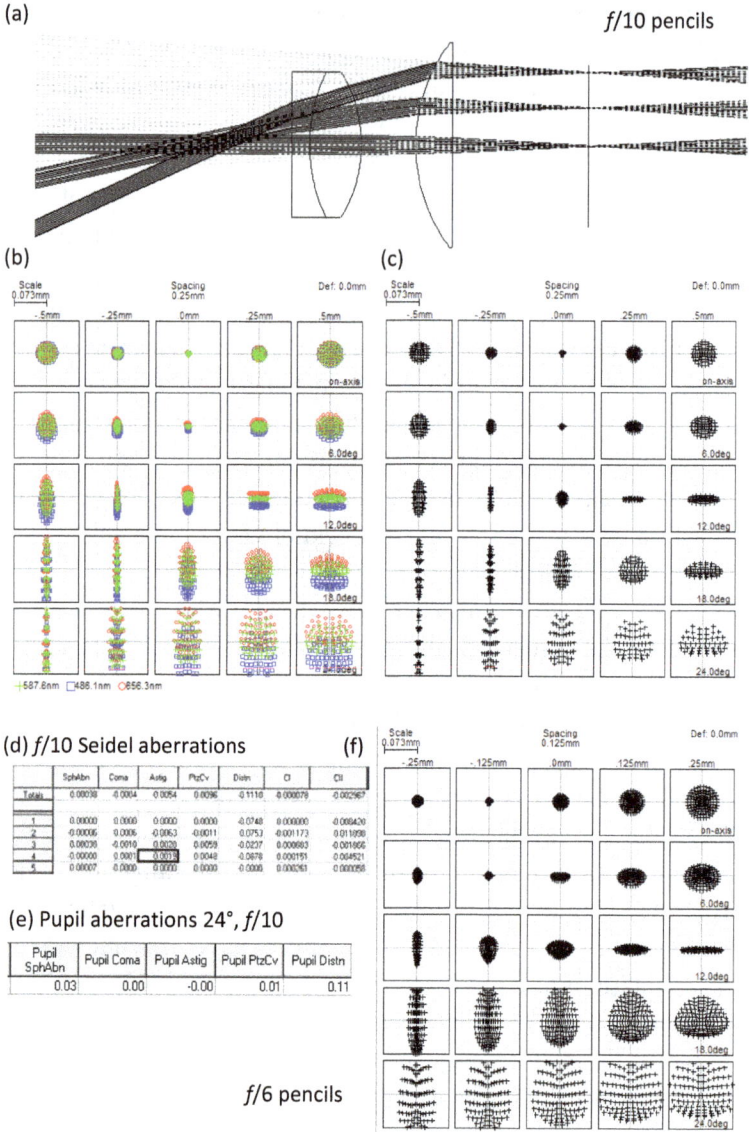

FIGURE 5.2 Kellner eyepiece by A. König 1953 [25 mm focal length, $F_{FL} = 22.7$ D, $F_{EL} = 19.7$ D, $d = 7.6$ mm]. (a) Ray traces for apparent field angles of 0°, 12° and 24°. (b) Spot diagrams for three wavelengths at apparent field angles of 0°, 6°, 12°, 18° and 24°. (c) Spot diagrams for $f/10$ pencils at apparent field angles of 0°, 6°, 12°, 18° and 24°. (d) $f/10$ Seidel wavefront aberrations for each surface and total (astigmatism S_{III} highlighted for surface 4, the convex surface of the field lens). (e) Pupil aberrations for eyepiece (24° field at $f/10$) (f) Spot diagrams for $f/6$ pencils at apparent field angles of 0°, 6°, 12°, 18° and 24°.

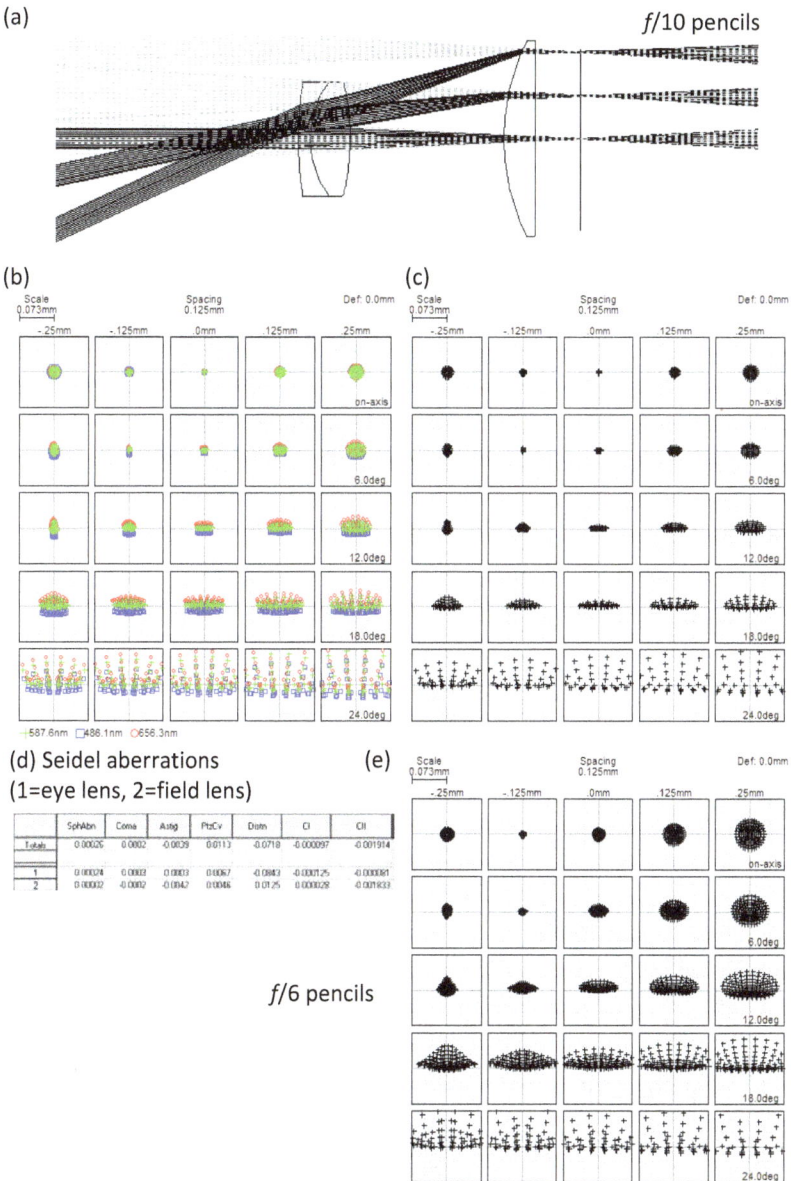

FIGURE 5.3 Kellner eyepiece by M. Kidger 2000 [25 mm focal length, $F_{FL} = 23.9$ D, $F_{EL} = 32.1$ D, $d = 19.4$ mm]. (a) Ray traces for apparent field angles of 0°, 12° and 24°. (b) Spot diagrams for three wavelengths at apparent field angles of 0°, 6°, 12°, 18° and 24°. (c) Spot diagrams for $f/10$ pencils at apparent field angles of 0°, 6°, 12°, 18° and 24°. (d) $f/10$ Seidel wavefront aberrations for each component (1 = eye lens, 2 = field lens) and total. (e) Spot diagrams for $f/6$ pencils at apparent field angles of 0°, 6°, 12°, 18° and 24°.

reduction in astigmatism is achieved by adding net positive astigmatism at the field lens: the surface entries in Figure 5.2d are changed to −0.0024 (cf. −0.0019) and +0.0025 (cf. 0.0000), with the latter change obviously contributing more.

It is debatable whether this improvement in astigmatism is useful or not, given the importance of also addressing residual chromatic aberration more successfully. In the same way, one could debate whether König's or Kidger's Kellner is better; perhaps they are just different, one offering better colour correction, and one offering better eye relief. For completeness, it should be noted that refraction by the Earth's atmosphere also introduces dispersion (chromatic separation of images), which may be particularly evident for objects observed at lower altitude and at higher magnification; not all dispersion evident through the eyepiece originates in the glass.[6]

Aberration Diagrams

So far, we have relied principally on spot diagrams and tables of Seidel wavefront coefficients to assess performance, but other diagnostics can be helpful, and we are now dealing with eyepiece designs of sufficient quality to justify looking more carefully at the aberrations. We briefly met longitudinal aberration plots in Figures 3.3, 3.4 and 3.7, and could benefit from re-examining these and other diagnostics before proceeding to evaluate other eyepiece designs.

Longitudinal aberration plots allow us to assess spherical aberration, coma, astigmatism and Petzval curvature. Figure 5.4 uses WinLens Basic ray-trace data for Kidger's Kellner eyepiece (Figure 5.3). The longitudinal aberration plots (Figure 5.4b) are shown in pairs, the left-hand column providing data in the meridional (tangential) plane, and the right-hand column providing data for the sagittal fan. The top pair of plots provides data for the pencil of rays coming from the on-axis object point, while subsequent pairs show pencils for off-axis rays, in this case at field angles of 6°, 12°, 18° and 24°. For clarity, a rectangle has been drawn around the longitudinal aberration plot for rays in the meridional plane for an on-axis object point; the other nine panels are drawn very closely to one another but can be distinguished with care.

The horizontal axis in each longitudinal aberration plot mimics the optical axis, which is usually defined as the z-axis, centred on the on-axis paraxial image point, so points to the left of centre lie to the left of the on-axis paraxial image point. The range covered by this axis can be set by the user but would usually cover a few millimetres. The vertical axis in

(a)

	SphAbn	Coma	Astig	PtzCv	Distn	CI	CII
Totals	0.00026	0.0002	-0.0039	0.0113	-0.0718	-0.000097	-0.001914
1	0.00024	0.0003	0.0003	0.0067	-0.0843	-0.000125	-0.000081
2	0.00002	-0.0002	-0.0042	0.0046	0.0125	0.000028	-0.001833

(b)

(c) (d)

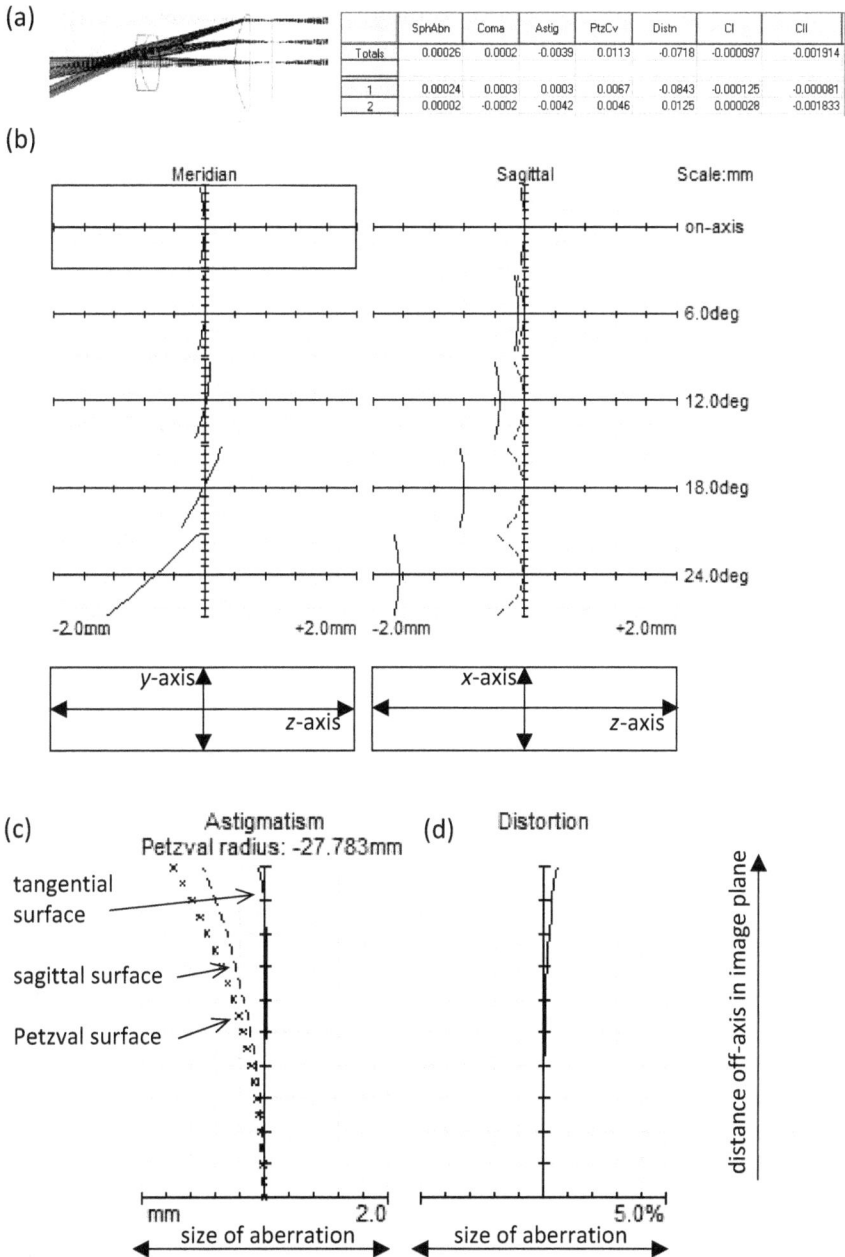

FIGURE 5.4 Aberrations of Kidger's 25 mm focal length Kellner eyepiece at f/10. (a) Lens layout and Seidel aberration coefficients at 24°. (b) Longitudinal aberration plots. (c) Field aberration plot for astigmatism. (d) Field aberration plot for distortion. See text for interpretation.

each panel corresponds to the centration at which each ray in the pencil passes through the aperture stop, ranging from the bottom to the top of the aperture for data in the meridional plane (defined already as our y-axis in Figure 3.1), and from the far side to the near side of the aperture for the sagittal fan. A schematic showing the orientation of the axes is included in Figure 5.4b.

Considering first the meridional ray data for an on-axis object point (top-left panel), the slightly curved arc shows where along the optical axis (z-axis) a ray at a given centration (y-axis value) crosses the principal ray for that pencil. The fact that the data lie along a shallow curve indicates they don't all cross the principal ray at the same point. As we saw in Section 3.6, this is spherical aberration, the only monochromatic aberration on-axis. As eyepiece lenses are rotationally symmetric about a common optical axis, which is a construction called a "centred system", the meridional rays and sagittal rays for an object point on the optical axis are the same, and hence there is no difference between the meridional and sagittal longitudinal aberration plots in the top row. However, for object points off the optical axis, the paths of sagittal rays and meridional rays are different, so the left panels (meridional rays) and right panels (sagittal rays) differ for off-axis points.

Recall that spherical aberration exists off-axis as well as on-axis, so the same shallow arc exists for the meridional rays and sagittal rays at the other off-axis points. However, a glance at the meridional ray plots off axis indicates that other aberrations are also coming into play, as the arc tilts progressively clockwise at greater field angles. This indicates that rays passing through the aperture stop above the principal ray ultimately intersect with it further to the right than rays passing through the aperture stop below the principal ray; this asymmetry is the signature of coma, and it is seen to increase steadily at high field angles. The longitudinal aberration plots for sagittal rays (right-hand column), meanwhile, do not show such an asymmetry, as sagittal rays passing behind and in front of the principal ray travel symmetric paths.

In addition to the shallow curve of spherical aberration and the tilt of the spherical aberration curve for meridional rays due to coma, we can assess astigmatism. The solid curves on the meridional plots are centred at the same z-axis distance as the paraxial on-axis ray, but the solid curves on the sagittal plots migrate progressively to the left at higher field angles. These two curves indicate that the meridional (tangential) rays come to a focus in the paraxial focal plane, so will form a horizontal focal line in the spot diagrams, but the sagittal focal surface lies ahead of that and will form a vertical focal line in the spot diagrams; this z-axis separation of the longitudinal aberration arcs for meridional and sagittal rays is astigmatism, and as expected the separation increases at higher field angles.

Astigmatism can perhaps more easily be assessed via a different diagram known as a field aberration plot (Figure 5.4c). This has different axes to the longitudinal aberration plot: the vertical axis represents the distance of image points from the optical axis, and the horizontal axis shows the size of the aberration at those various distances. The astigmatism plot contains three curves: the solid curve gives the position of the tangential focal surface, the dashed curve gives the position of the sagittal focal surface, and the cross symbols show the position of the Petzval surface. (It may be recalled from the discussion of Seidel S_{III} and S_{IV} coefficients in Section 3.7 that the distance of the Seidel tangential focal surface from the Petzval surface is three times the distance of the Seidel sagittal focal surface from the Petzval surface; this is clear from Figure 5.4c). A second field aberration plot is shown in Figure 5.4d, for distortion.[7] The field aberration plots provide more insight than the Seidel coefficients of the extreme field angle into how the aberrations deteriorate at higher field angles. The field aberration plots also illustrate more clearly whether higher-order (non-Seidel) aberrations are coming into play, which becomes evident if a field aberration curve departs from the quadratic or other h' dependence expected from Seidel aberration theory. The field aberration plots also reinforce the necessity of implementing a field stop to restrict the apparent field of view of an eyepiece, preventing poorly formed and distracting images from being presented to the eye.

The longitudinal aberration plots (Figure 5.4b) and field aberration plots (Figure 5.4c) indicate how the aberrations change for different ray paths through the aperture stop and for different field angles, respectively. It can also be instructive to look at transverse aberration plots, which indicate how much the rays are separated in the image plane (usually the plane containing the paraxial, on-axis image). This diagnostic also conveys the degradation of image quality, since ray separations in the image plane manifest themselves as a blurring of images. Of course, they show the same underlying aberrations, but present the data in a different way, which can be helpful. Transverse aberration plots (Figure 5.5b) are set out slightly differently from the longitudinal aberration plots. The horizontal axis now represents the location of the ray within the aperture stop, either the y-axis (meridional plane) position for the meridional plane aberrations

(a)

	SphAbn	Coma	Astig	PtzCv	Distn	CI	CII
Totals	0.00026	0.0002	-0.0039	0.0113	-0.0718	-0.000097	-0.001914
1	0.00024	0.0003	0.0003	0.0067	-0.0843	-0.000125	-0.000081
2	0.00002	-0.0002	-0.0042	0.0046	0.0125	0.000028	-0.001833

(b)

(c)

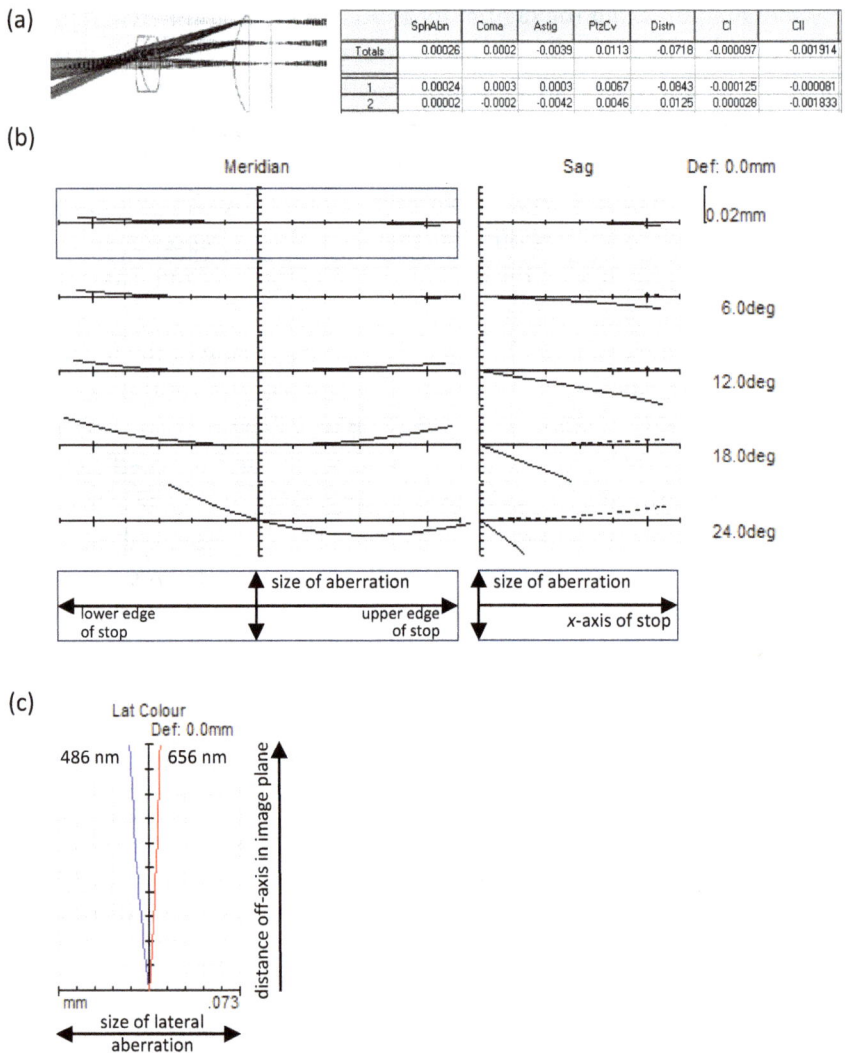

FIGURE 5.5 Transverse aberrations of Kidger's 25 mm focal length Kellner eyepiece at $f/10$. (a) Lens layout and Seidel aberration coefficients for 24° field. (b) Transverse aberration plots. (c) Transverse chromatic aberration plots.

(left-hand column), or the x-axis position for the sagittal ray aberrations (right-hand column). Ray paths having negative x-values are symmetric with those having positive x-values, so only the latter are shown. The vertical axis represents the size of the transverse aberration. For meridional rays, this is the displacement above or below the principal ray in the image plane. The displacement of sagittal rays is also relative to the principal ray, but sagittal rays can be displaced both in the horizontal and vertical directions in the image plane, hence there are two curves for sagittal rays at each field angle; the horizontal displacements tend to be greater than the vertical ones, so sometimes the vertical displacements are omitted for clarity.[8]

In the case of spherical aberration, which we recall is the only monochromatic aberration on-axis, meridional rays passing through the aperture stop at the *lower* edge will generally[9] intersect the principal ray before reaching the paraxial focal plane, so they will pass through the paraxial focal plane above the principal ray. Therefore, in transverse aberration plots, the on-axis, meridional ray panel will show points at the left-hand end having positive transverse aberrations. An on-axis point giving rise to meridional rays passing close to the *upper* edge of the aperture stop will likewise cross the principal ray prior to the paraxial focal plane and will therefore pass through the paraxial focal plane below the principal ray. Therefore, rays at the right-hand edge of the on-axis, meridional-ray transverse aberration panel will exhibit negative transverse aberrations. Consequently, spherical aberration manifests as a downward sloping trend in the on-axis meridional-ray transverse aberration panel (Figure 5.5b).[10] The aberration depends quadratically on ray centration, so the trend curves away from the horizontal line at the far left and far right, corresponding to the edges of the aperture stop. A purely linear trend in this diagram could be compensated by an appropriate shift of the preferred focal plane (a "defocus"), and a suitable defocus value to intercept the "circle of least confusion" (the narrowest waist of the pencil of rays) can indeed reduce the transverse aberration due to spherical aberration, but it cannot remove spherical aberration. It would therefore result in a tilting of the transverse aberration plot closer to the horizontal but without removing the curvature.[11]

We now consider the effect of coma on the meridional transverse aberration plots. The effect of positive coma is to cause meridional rays that pass close to the *upper* edge of the aperture stop to reach the paraxial focal plane further away from the optical axis than the paraxial ray, and thus it contributes positive transverse aberration. But with positive coma, meridional rays passing close to the *lower* edge of the aperture stop *also* reach the paraxial focal plane further away from the optical axis than the paraxial ray, so they also receive a contribution of positive transverse aberration. In other words, with positive coma, both ends of the off-axis meridional-ray transverse aberration plots are bowed upwards.[12] This signature of coma is very clear at the 12°, 18° and 24° field angles in Figure 5.5b. Recall that it affects only off-axis field angles, and that Seidel coma increases linearly with field angle.

If the transverse aberration curves for meridional rays and sagittal rays slope in opposite directions, for example, the meridional ray transverse aberrations become more negative with increasing centration while sagittal ray aberrations become more positive, then this is a sign of astigmatism.[13,14] Such a behaviour is evident in the 12° and 18° transverse aberration curves of Figure 5.5b, and consistent with Figure 5.4c.

It is worth noting again that the Seidel treatment of aberrations is still just an approximation, at the third order of the sin function (see Equation 2.4 and Section 3.5), and that at high field angles and at the extremes of the aperture stop, higher- (fifth-) order aberrations may come into play. These may be implicated by more dramatic excursions of the field aberration and transverse aberration curves from the behaviour expected on the basis of the Seidel analysis.[15]

Another transverse aberration diagram shows the lateral chromatic aberration, as in Figure 5.5c.

Diversity of Designs

For completeness, we note that designs for Kellner eyepieces possessing a three-element eye lens (instead of the conventional two) also exist, but we have refrained from presenting their details here. They may be investigated separately by the reader if sufficiently motivated.[16] In another variation, Heinrich Erfle designed[17] a four-element eye lens, assembled as two cemented doublets, which could be used to achromatise *and* increase the field of view of both Huygens and Ramsden eyepiece types up to 70°. However, Erfle's name is more commonly synonymous with a more fundamental redesign of eyepieces, which we investigate in Section 5.6.

The diversity of possible forms for the Kellner eye lens is a reminder of a more general fact that there is a wide range of eyepiece designs, each of which departs in some fashion from those preceding it. There is no strict definition of when a given eyepiece design ceases to belong to an existing type and should be distinguished by a new name. In this chapter in particular, we meet numerous examples of such diversity. For this book, I have taken the view that if the layout of the lens groups is still recognisable, then the "mere" bending, splitting or combining of elements or components to make minor adjustments to the aberrations is generally still consistent with the pre-existing type. However, the introduction of a distinctive new group, or a very different looking set of elements within a group, would probably be sufficient to justify a new type. In astronomy, the type is typically named after the originator, so the section just concluded has described all eyepieces comprising two groups where the field lens group is single and where the eye lens group comprises two to four achromatising elements, as a Kellner eyepiece, irrespective of where Carl Kellner himself ever imagined such modifications. The next four sections also consider two-group eyepieces, but where the field lens is no longer single, and where the distribution of elements within the groups is also dissimilar, and they are reasonably known by different names. This is a reasonable convention, but it is not an immutable law of optics. Additionally, we should note that most patents include terminology such as "the specification and the drawing are to be interpreted as illustrative and not in a limiting sense"[18]; in other words, making changes to an existing design will not necessarily mean it escapes designation as the original designer's type, even if that designer did not explicitly set out or even contemplate the later change.

One commercial example of a Kellner eyepiece with a moderately clear description of the adopted form is the Celestron® E-Lux® series, which spans focal lengths 26–40 mm in a 2 inch eyepiece barrel, and which is specified as a 3-element, 2-group Keller, offering a 56° apparent field of view.[19] However, a practical consequence of the diversity of designs is that one manufacturer's eyepiece of a given focal length and type will quite likely not be the same as another's, and a manufacturer's designs may be altered over time without purchasers being aware. The purchaser of a "25 mm Kellner", for example, is unlikely to have much, if any, information on the specific optical design that has been employed in the particular eyepiece they buy, which is a less-than-ideal state of affairs for the buyer. A very clear, and in my opinion commendable, exception is the line of Edmund Optics® RKE® eyepieces, which we examine in Section 5.2, for which Edmund Optics® publishes very detailed optical specifications on their website.[20] In contrast, the so-called "modified achromat (MA)" eyepiece is an obvious example of an undefined design, where the terminology suggests it is a derivative of the Kellner, but how it has been modified is usually not stated, and whether one manufacturer's MA bears any resemblance to another's is similarly unclear. We will see similar vague terminology applied to the classic Plössl design in Section 5.5, where "modified Plössl" and "Super Plössl" are uninformative but commonly marketed names.

The discussion of modifications to the Kellner eyepiece also indicates there is scope to modify designs more broadly to improve performance and reduce aberrations. We discuss inverted/reversed forms of the Kellner next, then the orthoscopic eyepiece as a different approach to an achromatised field lens paired with a simple, single eye lens. We shall then meet the König and Plössl forms that depart further from the Huygens-Ramsden-Kellner sequence of two-group eyepieces.

5.2 INVERTED/REVERSED KELLNER EYEPIECE

The so-called "inverted Kellner" or "reversed Kellner" eyepiece has the field lens achromatised while the eye lens is a single lens.[21] One eyepiece fitting this description, designed by Ludwig Bertele,[22] consists of a cemented doublet field lens with a leading, negative flint element to which is bonded a trailing positive lens, followed by a single positive eye lens. The patent also notes that the field lens could comprise three elements, and the eye lens two; a detailed specification for a three-element field lens is given, which resembles an orthoscopic eyepiece (Section 5.3). Bertele suggests a field of view of 60°–70°, entering a new regime compared to the eyepieces we have discussed so far. Bertele's reversed Kellner (Figure 5.6) provides better monochromatic images over a wider field than Kidger's Kellner, so achieves some of this ambition, but with worse chromatic aberration. However, Bertele also achieves an impressive 18 mm eye relief, significantly more than Kidger's 10 mm, making the reverse Kellner a potentially attractive eyepiece.

A related design, which was commercialised and is advertised with an apparent field of view of 45°, is the Edmund Optics® RKE® design.[23] The layout, detailed in optical design files on the company's website, is very similar to Bertele's, but with the negative flint bent to present a leading concave surface in a modification attributed to David Rank, and adopting symmetric forms for the two biconvex elements, which simplifies manufacture. Rank's obituary[24] describes the RKE® eyepiece as a computer-optimised Type II Kellner, designed after his 1974 retirement from Penn. State University while consulting for Edmund Optics.

The current (2025) RKE® range includes focal lengths of 21.5 and 28 mm. In Figure 5.7, I have rescaled their design to a focal length of 25 mm for comparison with the other 25 mm focal-length designs in this book and have set the lens diameter large enough to examine the field out to 24°. The design shown in Figure 5.7 is therefore *not* an Edmund Optics® RKE® eyepiece but closely follows their design principles and should demonstrate the general capabilities of this type of eyepiece on a similar basis to the other designs in this book.

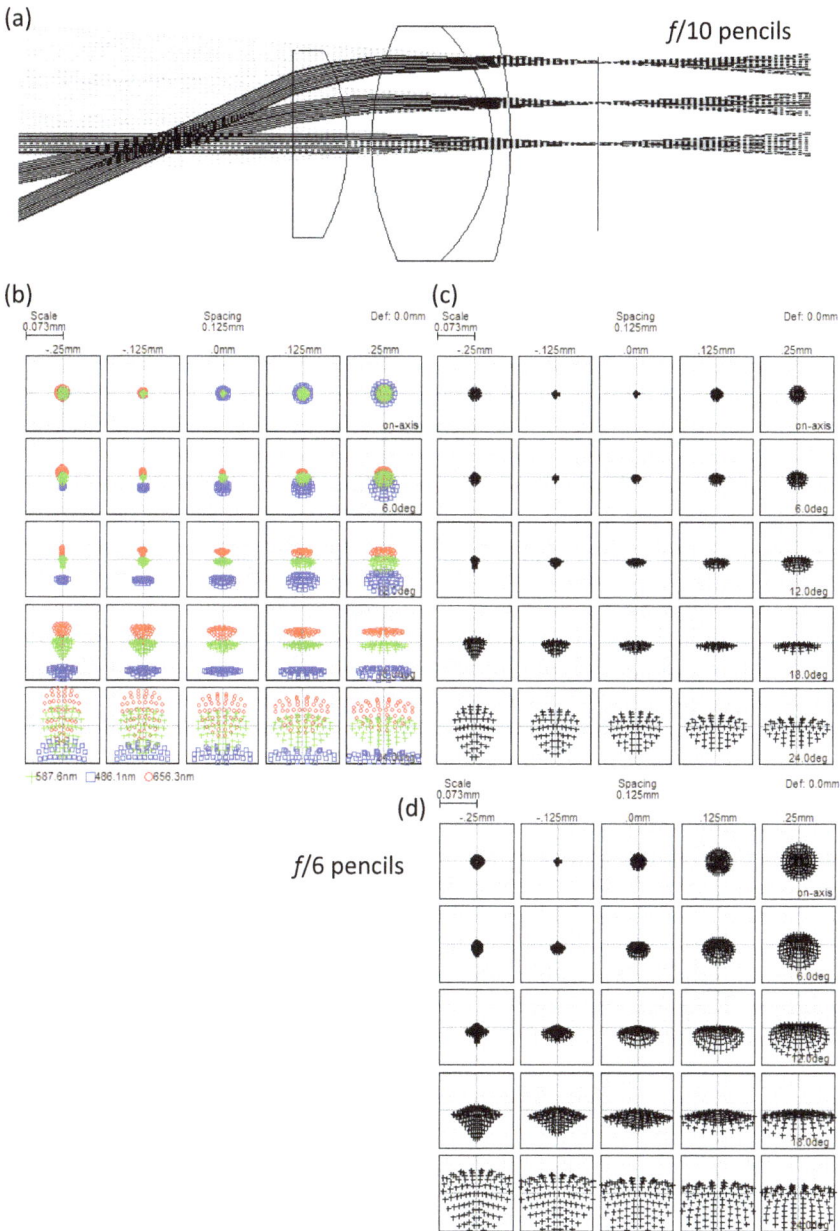

FIGURE 5.6 Reversed Kellner eyepiece by H. Bertele 1933 [25 mm focal length, F_{FL} = 17.1 D, F_{EL} = 25.6 D, d = 3.2 mm]. (a) Ray traces for apparent field angles of 0°, 12° and 24°. (b) Spot diagrams for three wavelengths at apparent field angles of 0°, 6°, 12°, 18° and 24°. (c) Spot diagrams for f/10 pencils at apparent field angles of 0°, 6°, 12°, 18° and 24°. (d) Spot diagrams for f/6 pencils at apparent field angles of 0°, 6°, 12°, 18° and 24°.

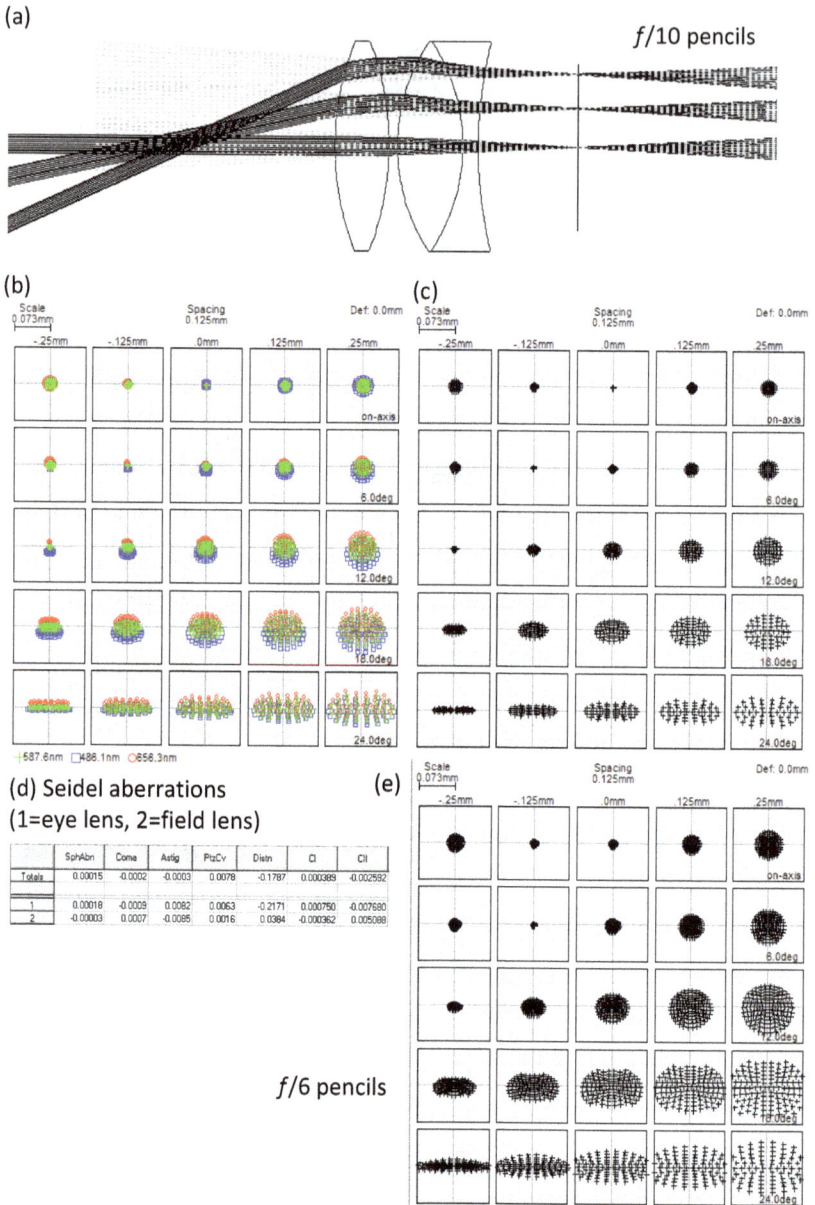

FIGURE 5.7 Reversed Kellner eyepiece rescaled from similar Edmund Optics RKE® design [25 mm focal length, F_{FL} = 6.7 D, F_{EL} = 30.1 D, d = 3.2 mm]. (a) Ray traces for apparent field angles of 0°, 12° and 24°. (b) Spot diagrams for three wavelengths at apparent field angles of 0°, 6°, 12°, 18° and 24°. (c) Spot diagrams for f/10 pencils at apparent field angles of 0°, 6°, 12°, 18° and 24°. (d) Seidel wavefront aberrations by component for 24° field at f/10. (e) Spot diagrams for f/6 pencils at apparent field angles of 0°, 6°, 12°, 18° and 24°.

The field lens in Figure 5.7 is notably weaker (6.7 D) than Bertele's in Figure 5.6 (17.1 D), and this helps achieve a longer eye relief, 25 mm. Moreover, the images are much rounder due to astigmatism being all but eliminated in the RKE®-like reverse Kellner design, which achieves $S_{III} = -0.0003$ mm (cf. -0.0028 in Bertele's). The lateral chromatic correction is also much improved, achieving $C_{II} = -0.0026$ mm (cf. -0.0060), as is the overall image quality in the 24° field. Field curvature is evident, but with astigmatism much reduced, accommodation by the eye will attempt to address this change as the eye scans the field (though see Section 3.7).

With such a significant improvement in design, it is instructive to compare this reverse Kellner (Figure 5.7) with a good traditional Kellner design (for which we adopt Kidger's Kellner, Figure 5.3). For the 24° field at $f/10$, Kidger's Kellner design provides better colour correction (Figure 5.3b and d) and less distortion (6%; cf. 16%), but the reverse Kellner addresses astigmatism better (Figure 5.7c and d) giving it a greater useable field of view, and it provides far greater eye relief (25 mm; cf. 10mm). The pupil spherical aberration is also much lower in the reverse Kellner design, viz. 0.03 mm (cf. 0.11 mm for Kidger's Kellner).

5.3 ORTHOSCOPIC/ABBE EYEPIECE

The orthoscopic eyepiece, designed by Ernst Abbe in 1860 and sometimes known by his name, is a considerable advance on historical designs. Its name derives from the Greek for a straight or correct view and is intended to convey the notion of minimal distortion, though, like all marketing claims, this should be quantified. It is also associated with a wider apparent field of view and longer eye relief than contemporaneous eyepieces, which also means it can be used at the shorter focal lengths necessary to achieve higher magnification. In view of this, the orthoscopic became the eyepiece of choice for planetary observing and double stars during the mid-to-late 20th century, though another eyepiece of the same generation – the Plössl (Section 5.5) – gained in popularity in the late 20th century. Nevertheless, the properties of the orthoscopic still make it a competitive choice for planetary and double star observers.

The orthoscopic has a triplet field lens followed by a single eye lens. In order to achieve long eye relief, the Carl Zeiss design[25] in Figure 5.8 emphasises the importance of keeping the separation of the field-lens and eye-lens small, in their case to $< f'_E / 3$, and reducing the thickness of the eye lens by using a single positive lens, which therefore reduces the distance between the principal planes and the rear vertex, thus

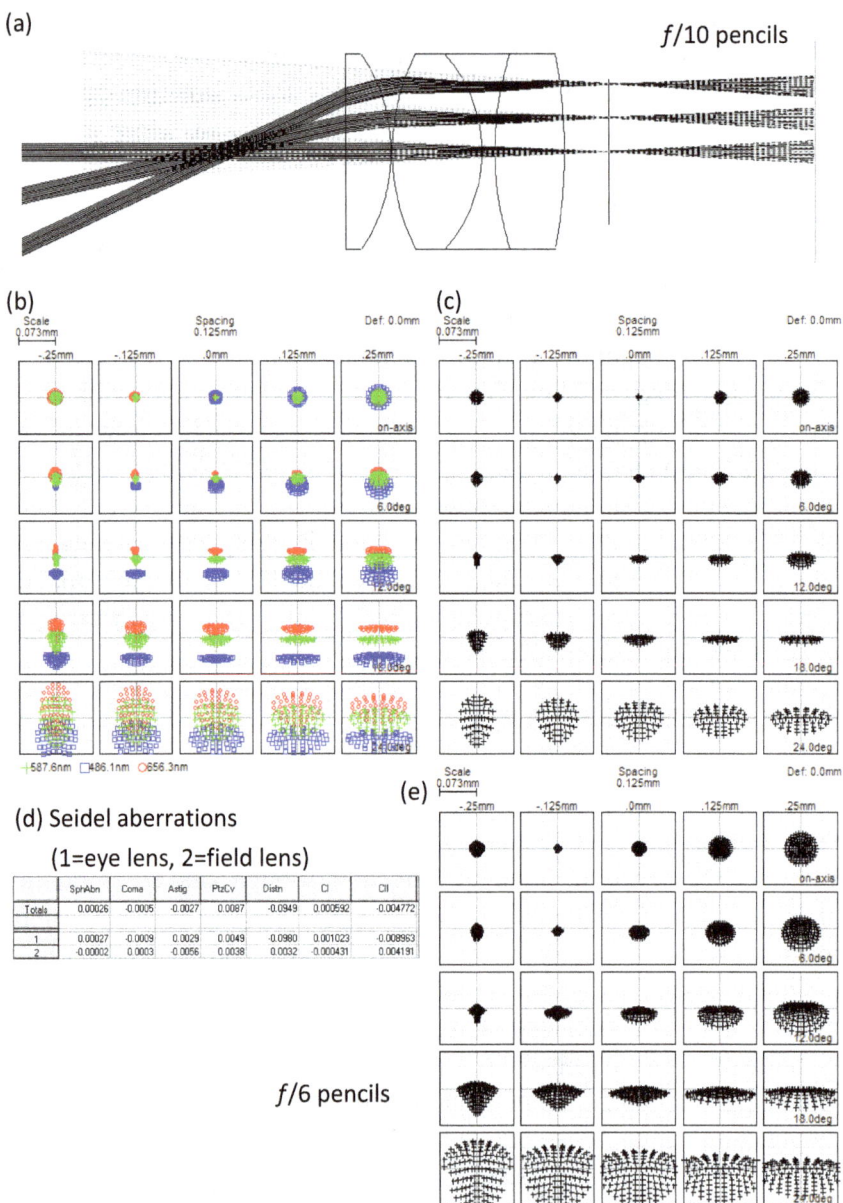

(a) $f/10$ pencils

(b)

(c)

(d) Seidel aberrations

(1=eye lens, 2=field lens)

	SphAbn	Coma	Astig	PtzCv	Distn	CI	CII
Totals	0.00026	-0.0005	-0.0027	0.0087	-0.0949	0.000592	-0.004772
1	0.00027	-0.0009	0.0029	0.0049	-0.0980	0.001023	-0.008963
2	-0.00002	0.0003	-0.0056	0.0038	0.0032	-0.000431	0.004191

(e)

$f/6$ pencils

FIGURE 5.8   Orthoscopic eyepiece for Carl Zeiss 1939 [25 mm focal length, F_{FL} = 14.4 D, F_{EL} = 26.5 D, d = 0.2 mm]. (a) Ray traces for apparent field angles of 0°, 12° and 24°. (b) Spot diagrams for three wavelengths at apparent field angles of 0°, 6°, 12°, 18° and 24°. (c) Spot diagrams for $f/10$ pencils at apparent field angles of 0°, 6°, 12°, 18° and 24°. (d) Seidel wavefront aberrations ($f/10$, 24°) by component. (e) Spot diagrams for $f/6$ pencils at apparent field angles of 0°, 6°, 12°, 18° and 24°.

increasing the second-vertex focal length and eye relief. Their design attains an eye relief of 20 mm, which clearly improves upon the results we saw previously for Huygens (5 mm), Ramsden (8–12 mm) and Kellner (10–14 mm) eyepieces of the same 25 mm focal length, though doesn't quite match the reverse Kellner (25 mm). The Zeiss design statement also emphasises the importance of high refractive index glasses so that the average of the refractive indices in the triplet field lens is greater than 1.6. This prompted the use of so-called dense crowns with $n = 1.57$ for the positive elements of the triplet, rather than a traditional borosilicate crown BK7 ($n = 1.52$). It is worth noting that historically, the wide varieties of optical glasses available now only appeared after the 1880s, with the development of a three-way collaboration between Otto Schott, Ernst Abbe and the Carl Zeiss company; the developers of the original Huygens, Ramsden and Kellner eyepieces had a much more limited choice of glasses.

Has the eyepiece delivered improved image quality? The 18° field certainly provides better monochromatic image quality than the Kellner eyepieces, illustrating the slightly larger fields of view that the orthoscopic eyepieces can acquire. The field stops of orthoscopic eyepieces are typically set to admit apparent fields of view of around 40°, though at least one manufacturer[26] has extended this to 52° because of the practical benefit to the observer of having sight of their target of interest in the periphery of the field even if image quality is deteriorating, as an aid to locating and subsequently centring the target.

The orthoscopic eyepiece in Figure 5.8 still exhibits astigmatism at high fields, but less so than the Kellner eyepieces; the orthoscopic S_{III} value has diminished to −0.0027 mm; cf. −0.0054 for König's Kellner and −0.0039 for Kidger's Kellner. The other aberrations are also reduced or not significantly worse in the orthoscopic, except for the lateral colour $C_{II} = -0.0047$ mm (cf. −0.0030 for König's Kellner and −0.0019 for Kidger's Kellner). The difficulty of addressing chromatic aberration with a four-lens, orthoscopic eyepiece makes Huygen's "trick" of nulling the lateral chromatic aberration in his 1660s eyepiece using two lenses of the same glass all the more impressive!

The orthoscopic eyepiece is frequently described as being free of distortion. The Carl Zeiss example is certainly better (−8.5%) than König's Kellner (−9.8%), but Kidger's Kellner achieves −6.4%, so the "distortion-free" description should be viewed with caution nowadays as other eyepieces can achieve similar values. In practice, it is rare for astronomical observers to complain about distortion in their eyepieces; complaints about astigmatism or chromatic aberration are more common.

It makes sense to explore Bertele's eyepiece designs[27] of the same decade (1933, cf. 1939 for the Carl Zeiss orthoscopic), which we briefly noted earlier (Section 5.2), since they share the concept of a composite field lens and simple eye lens. Bertele's eyepiece with a two-element field lens produces only slightly inferior performance compared to the Carl Zeiss orthoscopic, while Bertele's second eyepiece (Figure 5.9), which having a three-element field lens resembles the Carl Zeiss orthoscopic, performs even better. The useable field angle pushes out to 24° with significantly diminished astigmatism (S_{III} = −0.0019 mm), although the lateral colour is slightly worse: C_{II} = −0.0063 mm. The eye relief is a respectable 17–18 mm.

An important feature of the ray paths for the RKE and orthoscopic eyepieces (Figures 5.7a, 5.8a and 5.9a) is that the rays exiting the field lens towards the eye lens do so at almost the same height as they entered the field lens. This places the principal plane of the eyepiece close to the eye lens, which is achieved in part by the first element of the field lens being a negative lens with a high refractive index (a strong flint), followed immediately by a much lower refractive index, positive lens (Bertele suggests a difference $\Delta n > 0.15$). This is crucial to delivering a long eye relief, as it keeps the principal rays away from the optical axis until they are almost ready to encounter the final, single-element eye lens. As the eye lens is quite simple and thus has very limited aberration control, the multi-element field lens has the role of preconditioning the rays to pre-empt the aberrations that the eye lens will introduce and thereby limit their overall sum. The Seidel wavefront aberrations for Bertele's eyepiece are tabulated in Figure 5.9d, split into (1) the eye-lens and (2) field-lens contributions. In most instances, the aberrations from the two components have opposite signs so that they counteract one another where possible, though it is not possible for two positive components to have opposite Petzval curvature, of course. The pupil spherical aberration term is also low, just 0.03 mm and 0.04 mm for the Zeiss and Bertele eyepiece, respectively.

Orthoscopic eyepieces are currently made by a relatively small number of manufacturers including Baader[28] (6–18 mm focal length) with a 52° apparent field of view, Masuyama[29] (4–25 mm focal length) with the classical 42° field, Omegon[30] (10.5–24 mm focal length) with 40°–46° apparent field, and Takahashi[31] (4–32 mm focal length, 44° apparent field) which market it as the Abbe; all provide eye relief slightly shorter than the focal length.

(a)

$f/10$ pencils

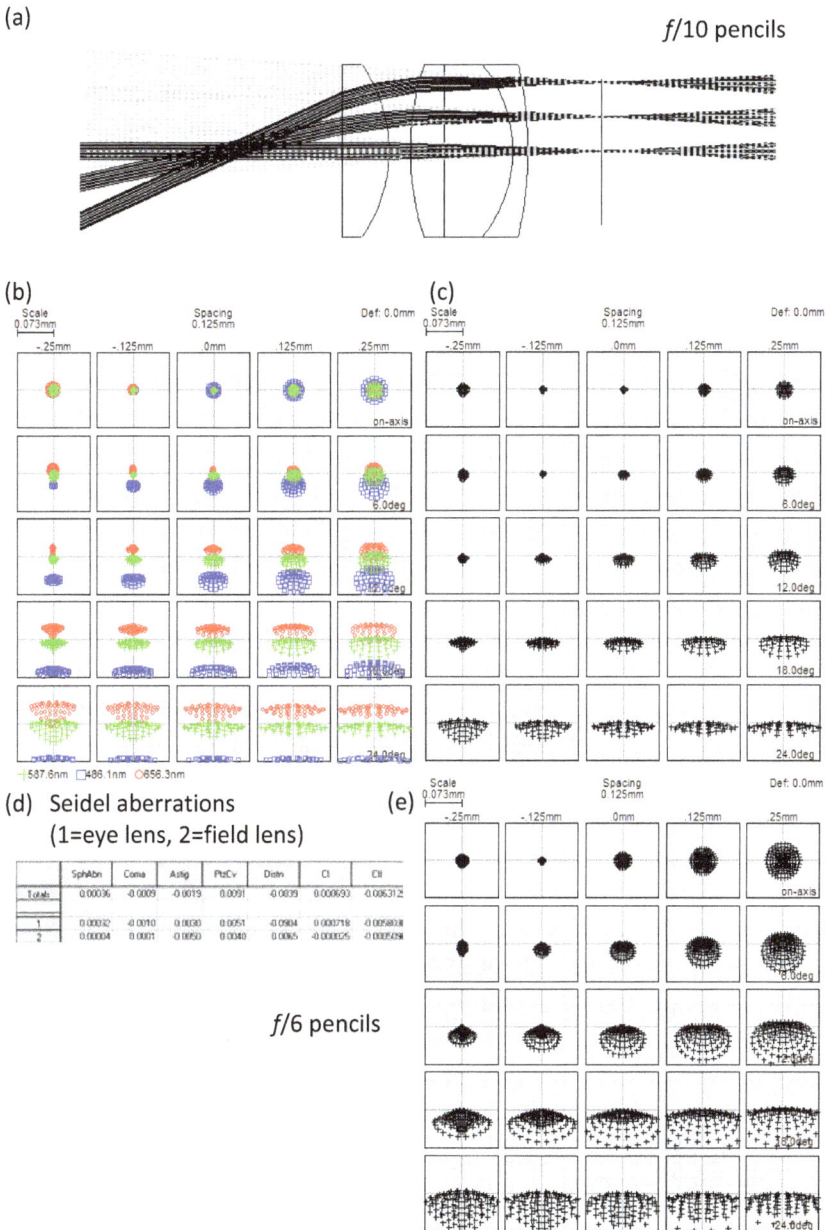

(b)

(c)

(d) Seidel aberrations
(1=eye lens, 2=field lens)

	SphAbn	Coma	Astig	PtzCv	Distn	CI	CII
Totals	0.00036	-0.0009	-0.0019	0.0091	-0.0039	0.000693	-0.006312
1	0.00032	-0.0010	0.0030	0.0051	-0.0904	0.000718	-0.005803
2	0.00004	0.0001	-0.0050	0.0040	0.0065	-0.000025	-0.003428

(e)

$f/6$ pencils

FIGURE 5.9 Bertele's 1933 orthoscopic eyepiece [25 mm focal length, $F_{FL} = 17.1$ D, $F_{EL} = 25.6$ D, $d = 3.2$ mm]. (a) Ray traces for apparent field angles of 0°, 12° and 24°. (b) Spot diagrams for three wavelengths at apparent field angles of 0°, 6°, 12°, 18° and 24°. (c) Spot diagrams for $f/10$ pencils at apparent field angles of 0°, 6°, 12°, 18° and 24°. (d) Seidel wavefront aberrations ($f/10$, 24°) by component. (e) Spot diagrams for $f/6$ pencils at apparent field angles of 0°, 6°, 12°, 18° and 24°.

5.4 KÖNIG EYEPIECES

There is no doubt that Albert König was a prolific designer! Besides the 1953 Kellner eyepiece design[32] discussed in Section 5.1 and a much earlier (1907) Kellner variant using a three-element eye lens,[33] König produced four designs of 3–4 element, 2-group eyepieces (1915),[34] three of which resemble an inverted Kellner with a single eye lens and multi-element field group. A three-element design was marketed under König's name[35] as a moderately wide-field eyepiece in the later decades of the 20th century,[36] though there is some uncertainty over precisely which of his designs was/ were employed or whether they were modified further,[37] one description calling them a modified Erfle,[38] and another a simplified orthoscopic reflecting the two-group, single-eye-lens designs. The fourth design of the 1915 patent resembles a Plössl eyepiece, a type we consider in the next section. Wider field eyepieces designed by König are also described in Section 5.6.

5.5 PLÖSSL EYEPIECE

While an orthoscopic eyepiece shares some features with an inverted Kellner, i.e., an achromatised field lens followed by a single eye lens, the Plössl displays a more symmetric form, comprising two doublets (often plano-convex but not necessarily so) with outward-facing flints, where the two components are generally very similar to one another, if not identical. Outside astronomical circles, the Plössl is known alternatively as a "symmetric" or sometimes "military" eyepiece.[39] The origin of the last of these names is evident in a description[40] of the quest for greater eye relief in eyepieces used in the sights of guns, where recoil of the gun when discharged could damage the eye of the observer. (Astronomical observers who have inadvertently ventured too close and too forcefully to an eyepiece in the dark may also appreciate that danger.) The Kellner eyepiece specifically was identified as unsatisfactory in that description, while the better-suited Plössl was referred to under another of its alternative names: "dial-sight eye piece".

The apparent symmetry of the lenses in a Plössl can be misleading. Symmetry in telescope objective lenses such as the Cooke triplet,[41] and many early photographic lenses such as the periscopic, rapid rectilinear, Ross,[42] Dagor, Celor and double Gauss,[43] where lenses display approximate front-back symmetry about a central aperture stop, is regarded as a good way of controlling coma, distortion and lateral colour.[44] However, the ray paths in an eyepiece are not symmetric about an aperture stop, as the entrance and exit pupils lie outside the eyepiece, asymmetrically. So, the symmetry of the lenses in a Plössl eyepiece does not have the same significance as it does in those other applications.

The design of the Plössl eyepiece can perhaps be better understood not as an evolution of the Huygens/Ramsden/Kellner/orthoscopic family, but by starting again with the simple, single-lens eyepiece that we investigated back in Chapter 3. That eyepiece was doomed to failure because of major off-axis aberrations. One means of dealing with aberrations from strongly powered lenses is to split the power into two separate lenses bearing approximately half the power each. This weakens each surface by around a factor of two, which considerably reduces most aberrations except the Petzval curvature, which even for a split lens system still depends largely on the sum of powers. With the spacing of the two Plössl components being minimal, and the first component no longer being particularly near the focal plane of the telescope, it no longer serves the same function that the field lens did in Chapter 4. Nevertheless, we will retain the field-lens and eye-lens terminology to describe the first and second component in the assembly since this terminology is well established and works well enough for the Plössl regardless.

By splitting the power of a Plössl more or less equally across two components that are almost in contact, we can also make use of another general result. If a system of two optical components is to be rendered free of longitudinal and lateral chromatic aberration, then it requires either that both components are individually achromatic, or they are in contact.[45] The first condition helps explain why none of the eyepieces we have reviewed so far have been capable of eliminating all chromatic aberration, as the Huygens and Ramsden eyepieces considered in Chapter 4 comprised only single, non-achromatised components (the Huygens eyepiece did at least correct lateral but not longitudinal chromatic aberration), and the Kellner, inverted Kellner and orthoscopic eyepieces discussed so far in Sections 5.1 and 5.3 contain either a single field lens or a single eye lens. Could splitting the eyepiece into a twin-doublet Plössl configuration finally solve the problem of chromatic aberration?

The Carl Zeiss patent[46] that we met in the preceding section on orthoscopic eyepieces also contains a design for a Plössl eyepiece, shown in Figure 5.10. The spot diagrams show a considerable further improvement in the off-axis image quality, with smaller images and better lateral colour correction. Surprisingly perhaps given the improvement in the monochromatic spot diagrams, the Seidel aberrations are almost identical for both Zeiss eyepieces, and even the

(a)

f/10 pencils

(b)

Scale 0.073mm   Spacing 0.125mm   Def: 0.0mm

-.25mm   -.125mm   .0mm   .125mm   .25mm

on-axis

6.0deg

12.0deg

18.0deg

24.0deg

+587.6nm □486.1nm ○656.3nm

(c)

Scale 0.073mm   Spacing 0.125mm   Def: 0.0mm

-.25mm   -.125mm   .0mm   .125mm   .25mm

on-axis

6.0deg

12.0deg

18.0deg

24.0deg

(d) Seidel aberrations
(1=eye lens, 2=field lens)

	SphAbn	Coma	Astig	PtzCv	Distn	CI	CII
Totals	0.00029	0.0006	0.0026	0.0086	-0.0964	0.000251	0.003295
1	0.00023	0.0007	0.0018	0.0045	-0.0842	0.000280	-0.003307
2	0.00006	0.0001	0.0044	0.0041	-0.0141	0.000029	0.000012

(e)

Scale 0.073mm   Spacing 0.125mm   Def: 0.0mm

-.25mm   -.125mm   .0mm   .125mm   .25mm

on-axis

6.0deg

12.0deg

18.0deg

24.0deg

f/6 pencils

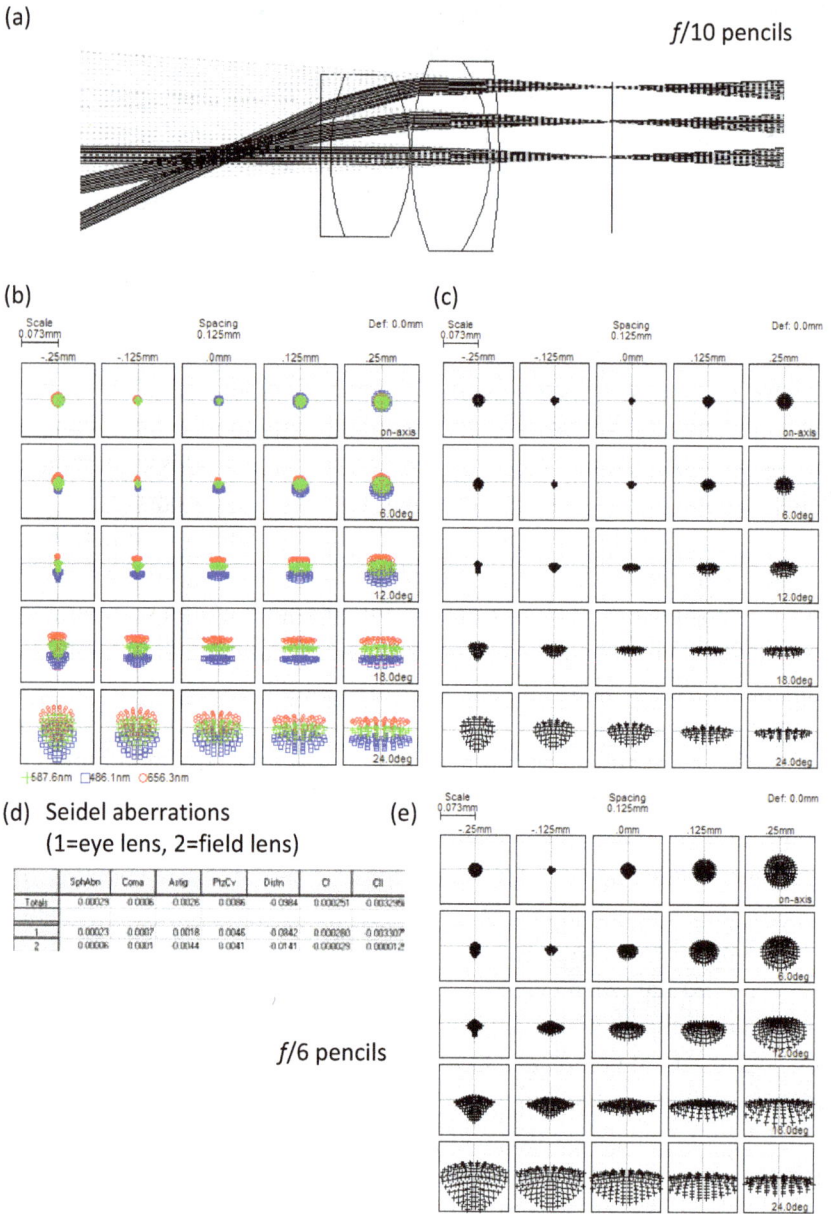

FIGURE 5.10 Plössl eyepiece for Carl Zeiss 1939 [25 mm focal length, $F_{FL} = 18.0$ D, $F_{EL} = 22.5$ D, $d = 0.15$ mm]. (a) Ray traces for apparent field angles of 0°, 12° and 24°. (b) Spot diagrams for three wavelengths at apparent field angles of 0°, 6°, 12°, 18° and 24°. (c) Spot diagrams for *f*/10 pencils at apparent field angles of 0°, 6°, 12°, 18° and 24°. (d) Seidel wavefront aberrations (*f*/10, 24°) by component. (e) Spot diagrams for *f*/6 pencils at apparent field angles of 0°, 6°, 12°, 18° and 24°.

astigmatism S_{III} is barely reduced in the Plössl (−0.0026 mm, cf. −0.0027 for the orthoscopic), but recall that the Seidel coefficients only quantify third order aberrations and provide no information on higher order aberrations (fifth order etc.) that also vary between designs and which will be evident in the ray traces and spot diagrams. The biggest improvement though is in the lateral colour C_{II}, which reduces to −0.0033 (cf. −0.0047); see Figure 5.10b and d. Being doublets, the two components of the Plössl eyepiece are each better corrected in the eye lens and field lens (C_{II} = −0.0033 and +0.0000) than in the orthoscopic components (cf. −0.0089 and +0.0042).

Amongst the 25 mm focal length eyepieces evaluated so far in this book, the Carl Zeiss Plössl is the clear winner on image quality. The improvement comes at the cost of the eye relief, which is reduced to 16 mm, compared to 20 mm for the Carl Zeiss orthoscopic. For a 25 mm focal length eyepiece, this difference is probably tolerable, but as the eye relief in these designs scales more or less linearly with focal length, the shorter focal-length eyepieces necessary for higher magnification incur progressively shorter eye reliefs, and an orthoscopic might still be preferred at short focal lengths even if that implies a smaller useable field of view.

The Zeiss Plössl has slightly different shapes for the four lenses that make the two components, though both components are ultimately plano-convex. Albert (Al) Nagler[47] and Michael Kidger[48] provide alternative Plössl designs adopting weakly concave rather than plano outer surfaces, where both components have matching radii of curvature (though the eye lens can be reduced in physical radius since the rays are less spread out). Nagler declared a willingness to suffer distortion and pupil aberrations in favour of other aspects of image quality more important to astronomical observers (see also Section 3.7), especially since the deterioration of image quality off-axis ultimately limits the field of view of an eyepiece. As Nagler's and Kidger's designs are quite similar, we show just one (Kidger's, as it happens) in Figure 5.11. Although its Seidel aberrations (Figure 5.11d) are mostly slightly larger than the Zeiss ones (resulting from plano-convex doublets; Figure 5.10d), the spot diagrams for Kidger's Plössl eyepiece (Figure 5.11b and c) are the best yet presented in this book. The off-axis images still exhibit some astigmatism, but the lateral colour correction is exceptional in comparison to all previously shown examples: C_{II} = −0.0024 (cf. −0.0033 for the Zeiss Plössl).

In addition to providing good images over a wider field of view than earlier eyepieces, and a reasonable eye relief at least in medium and long focal-length eyepieces, the Plössl has the advantage of relative simplicity,

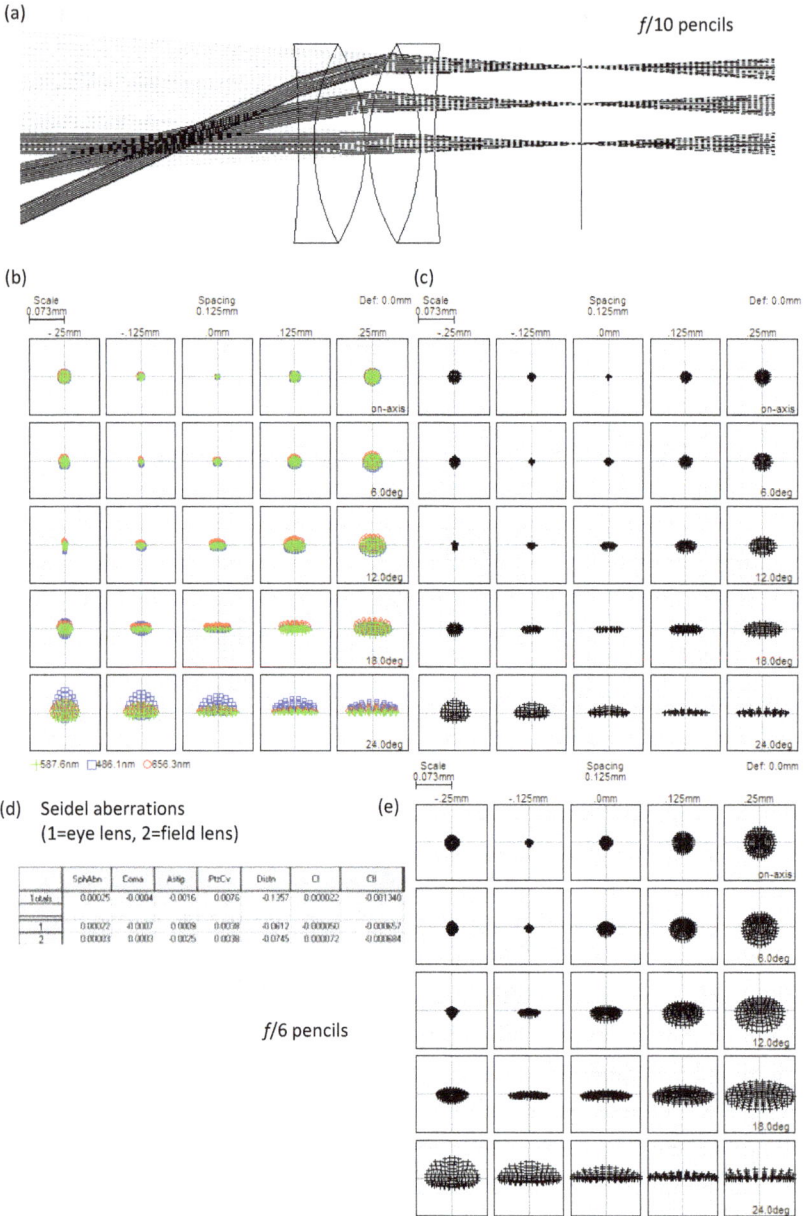

(a)

$f/10$ pencils

(b)

(c)

(d) Seidel aberrations
(1=eye lens, 2=field lens)

	SphAbn	Coma	Astig	PtzCv	Distn	CI	CII
Totals	0.00025	-0.0004	-0.0016	0.0076	-0.1257	0.000022	-0.001340
1	0.00022	-0.0007	0.0009	0.0038	-0.0612	-0.000050	-0.000657
2	0.00003	0.0003	-0.0025	0.0038	-0.0745	0.000072	-0.000684

$f/6$ pencils

(e)

FIGURE 5.11 Plössl eyepiece by M. Kidger 2000 [25 mm focal length, $F_{FL} = F_{EL}$ = 19.4 D, d = 0.5 mm]. (a) Ray traces for apparent field angles of 0°, 12° and 24°. (b) Spot diagrams for three wavelengths at apparent field angles of 0°, 6°, 12°, 18° and 24°. (c) Spot diagrams for $f/10$ pencils at apparent field angles of 0°, 6°, 12°, 18° and 24°. (d) Seidel wavefront aberrations ($f/10$, 24°) by component. (e) Spot diagrams for $f/6$ pencils at apparent field angles of 0°, 6°, 12°, 18° and 24°.

since the field lens and eye lens can be identical which reduces manufacturing costs, and with only four glass elements, the lens is of reasonable weight. It became a favourite among amateur astronomers towards the end of the 20th century, particularly for applications where a wide field of view was prioritised and only moderate magnification was desired, avoiding the short eye relief incurred at high magnification.

Current (2025) commercially available Plössls include (non-exhaustively) the following:

Antares® supplies[49] a range of Plössl eyepieces with 52° apparent field of view in 1¼ inch and, uncommonly, 1 inch barrels.

Baader® sells a 32 mm focal length Plossl.[50]

Bresser® markets[51] classical 50° apparent field of view, 4-element, 2-group Plössls in wide range of focal-lengths (5–40 mm), in 1¼ inch barrels (the 40 mm Plössl has a 46° field.)

The Celestron® Omni series is a 4-element, 2-group Plössl, with focal lengths from 4 to 40 mm in a 1¼ inch barrel and 56 mm in a 2 inch barrel, with apparent fields from 40° to 50°, and eye relief from 6 to 52 mm, scaling roughly with focal length.[52]

Omegon market a classic 4-element, 2-group Plössl[53] (4–40 mm focal length, 50° apparent field of view) and 5-element, 2-group Super Plössl[54] (4–56 mm focal length, 52° apparent field of view, except 46° for the 40 mm) eyepieces.

The Sky-Watcher® "SP" series is marketed as a Super Plössl, based on the "proven 4-element Plössl designs"[55]; focal lengths from 10 to 40 mm are available, giving an apparent field of view of 52°.

The Takahashi TPL range is described[56] as a modified Plössl, covering focal lengths 6–50 mm, conveying an apparent field of view of 48°, eye relief around 70% of the focal length, and with the image quality described as excellent in the inner 50% of the radius.

Tele Vue® also markets a wide range of Plössl eyepieces from 8 to 55 mm focal length, with 50° apparent field of view.[57]

The Vixen® NPL series[58] is a 4-element, 2-group Plössl spanning 4–40 mm focal lengths in a 1¼ inch barrel, and giving apparent fields of view of 50° (40° for the 40 mm); weights are 70–130 g. As usual for Plössl eyepieces, the eye relief in the NPL series is typically somewhat less than one focal length, which makes the eye relief on short-focal-length eyepieces very small.

Many astronomical equipment companies also supply Plössl and Super Plössl eyepieces bearing their own brand names, with the manufacturing undertaken by some other, usually unspecified, entity; the level of detail provided about such eyepieces varies. Uncertain details and origins can be a feature of longer established brands too, especially as the ownership of some historical brands has shifted away from their founders into larger holding companies owning multiple brands. As with many consumer goods, it is often unclear from marketing descriptions exactly what you are buying; I commend again the clarity provided by Edmund Optics® in publishing the detailed optical specification of their RKE® eyepiece.[59]

A long-established manufacturer of Plössl eyepieces missing from the list above is Meade®, which has marketed 6.4–40 mm focal length Series 4000 Super Plössls for many years, but as of January 2025, the inventory and selected intellectual property of Optronic, the holding company of Meade®, Orion® and Coronado®, has been sold,[60] and the future availability of these lines is uncertain.

The quest to extend better image quality to larger fields of view did not end with the Plössl, and designs for wider fields of view have existed since the early 20th century. One of the earliest of these to enter the mainstream was the Erfle eyepiece, which we discuss next, in Section 5.6, followed in Sections 5.7 and 5.8 by the quest for even greater apparent fields of view.

5.6 THREE-GROUP, WIDE-FIELD EYEPIECES

The Plössl eyepiece, as introduced above, can be viewed as a split version of the single-lens eyepiece (Chapter 3), where sharing the power across two components greatly reduces the total aberrations, and where each component is itself a cemented achromatic doublet with three surfaces and two glass types that permit chromatic aberration, coma and astigmatism to be addressed.[61] The vast improvement in performance that was achieved suggests that a further weakening of the components may prove useful, allowing a valuable increase in the tolerable field of view. The wide-field Erfle eyepieces, which we examine in the next subsection, implement this approach.

Erfle Eyepiece

Heinrich Erfle's 1925 designs[62] clearly begin from the Plössl form, with the intention of improving performance by separating the two doublets with a third, positive component. It is usually a single biconvex crown but potentially a doublet. The leading surface of the field-lens is concave, as in the RKE® reverse Kellner (Section 5.2) and Nagler's and Kidger's Plössl eyepieces (Section 5.5), but typically even stronger. This leading concave surface deflects the principal rays further away from the optical axis, providing the potential for a large eye relief when the principal rays are ultimately brought back to the optical axis.[63,64] The stronger deflection of the pencils away from the optical axis also results in Erfle eyepieces having a central cross-section that is wider than the field stop. (Plössl eyepieces may also have a slightly widened central cross-section if the upstream surface of the field lens is concave.) The inclusion of the positive lens in the middle of the eyepiece allows the facing convex surfaces of the doublets to be weakened. Figure 5.12 shows the second of the two designs in Erfle's patent. It has a 4.0 D field lens, 14.4 D central lens, and 18.3 D eye lens. Another design[65] has a 7.3 D field lens, 15.0 D central lens, and 13.2 D eye lens. The final eye-lens surface closest to the eye may be weakly concave, plano or weakly convex, depending on the details of the design. M. Kidger, for example, provides an Erfle design[66] in which that final surface is convex, but the final lens element is still a negative-power flint, so the layout does not differ substantially from that in Figure 5.12. A 1947 Erfle design by Fred Altman[67] adds an optional negative power field lens in the focal plane where it reduces the Petzval curvature without changing the focal length, and improves the pupil spherical aberration, but slightly worsens the astigmatism, and of course raises the risk of dust or scratches being in focus with the image.

The spot diagrams and Seidel aberration values in Figure 5.12c and d confirm that astigmatism continues to challenge the design, especially at high field angles. This is confirmed by the large S_{III} (astigmatism) value, −0.0042 mm. The chromatic performance (Figure 5.12b and d) for the 24°, $f/10$ pencil is good (C_{II} = −0.0056), though not as good as Kidger's Plössl eyepiece (Figure 5.11b and d; C_{II} = −0.0024). The pupil spherical aberration has also increased, to 0.05 mm (cf. 0.01 mm for Kidger's Plössl); this, combined with the large field of view of Erfle eyepieces, can require the observer to move their head forward to intercept the emergent beams corresponding to the edges of the field. The paraxial eye relief is 19 mm, which allows a little latitude for head movement, but a later variant by John Miles[68] had only 9 mm eye relief, so good eye relief is not a foregone conclusion in the Erfle design.

But is this Plössl vs Erfle comparison fair? Not really, as the Erfle eyepiece was designed to provide a wider field of view, out to field angles of 35°, i.e. a 70° field,[69] so comparing the eyepieces over the inner field only doesn't allow the Erfle to demonstrate its capabilities. So far in this book, the performance of all eyepieces has been compared on a similar basis,

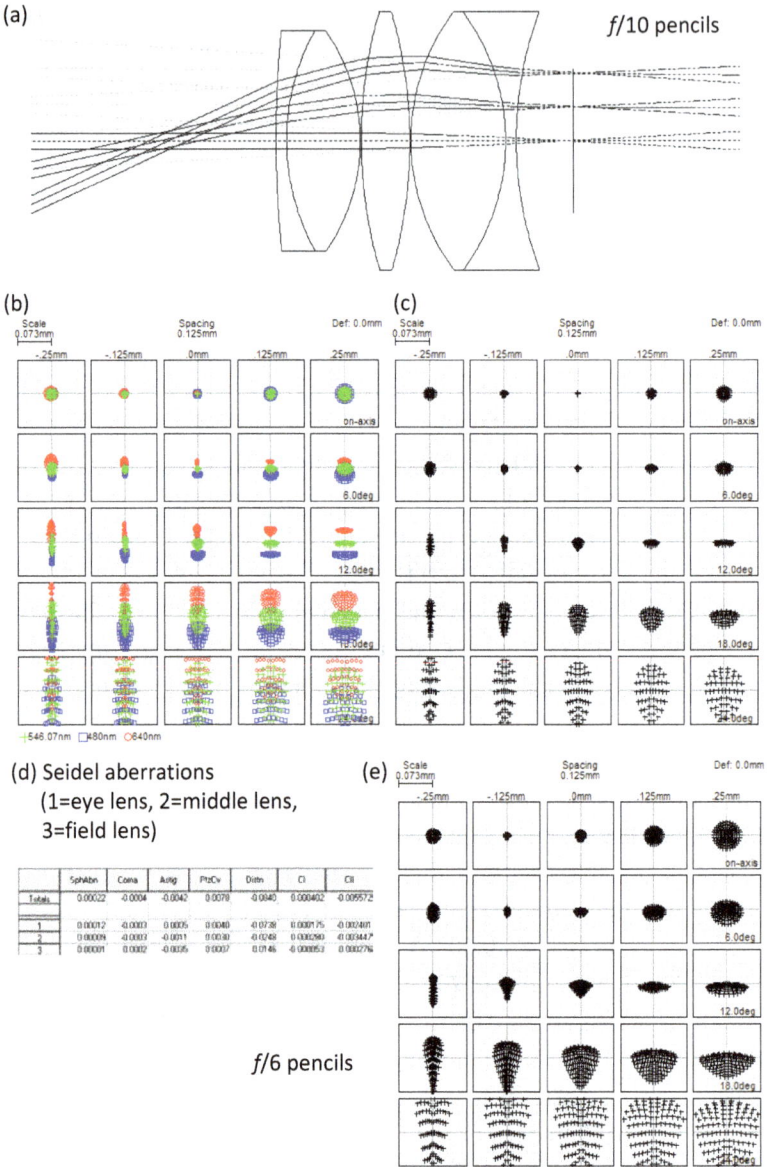

FIGURE 5.12 Erfle eyepiece [25 mm focal length, F_{FL} = 4.0 D, F_C = 14.4 D, F_{EL} = 18.3 D]. (a) Ray trace. The lens is larger than those shown previously because although the same field angles of 0°, 12° and 24° are shown, the eyepiece is designed to convey field angles up to 35°. (b) Spot diagrams for three wavelengths at apparent field angles of 0°, 6°, 12°, 18° and 24°. (c) Spot diagrams for f/10 pencils at apparent field angles of 0°, 6°, 12°, 18° and 24°. (d) Seidel wavefront aberrations (f/10, 24°) by component. (e) Spot diagrams for f/6 pencils at apparent field angles of 0°, 6°, 12°, 18° and 24°.

typically for fields of 0°, 6°, 12°, 18° and 24°, for $f/10$ and $f/6$ pencils, and with spot diagrams plotted to the same scale. As our discussion has now moved on to eyepieces designed to access fields of view of 60° or more, we need to compare wider field angles. A new comparison set of field angles 0°, 9°, 18°, 27° and 36° will now be calculated, starting again with Kidger's Plössl eyepiece since that was one of the best in our set out to 24°. The aberrations will inevitably deteriorate further off axis, so a new scale for the spot diagrams will also be adopted, compressing the previous scale by a factor of 2. It must be recognised that the Plössl was not intended to work at such wide fields, so any weakness in its performance in that regime is not a reflection on the capabilities of the designer; the comparison is purely to demonstrate why different designs are required to obtain good images at higher field angles. Spot diagrams for the Plössl and Erfle eyepieces of Figures 5.11 and 5.12, at $f/10$, are shown for the larger field angles in Figure 5.13a and b. Kidger's Plössl only admits fields out to 27° due to the sizing of the lenses. The Erfle is seen to maintain very good focus of the sagittal rays out to 36°, but obviously not for the meridional (tangential) rays, so stellar images would be considerably elongated, in this example pointing radially away from the optical axis. The decision on whether or not to use a wide-field eyepiece must be driven by need, since gaining field can come at the cost of pupil spherical aberration and increased astigmatism.

Further variants have been designed, with varying degrees of differentiation from Erfle's original design. König patented a wide-field eyepiece in 1940[70] presenting three examples, each of which comprises three components with a single, positive eye lens. The field and middle lenses comprise doublet + singlet, singlet + doublet, and triplet + doublet components respectively in the three designs. As in the Erfle, the splitting of powers across three components allows those components to be weakened, and therefore allows aberrations to be reduced, while at the same time keeping the lens train short and the eye lens single so that eye relief can be maximised. König's 1940 designs therefore share some similarities with the Erfle, but König's adherence to a simple, single eye lens and the avoidance of a strongly concave leading surface on the field lens gives his 1940 designs a different appearance.

König's 1940 design #1, like the Erfle, has a doublet field lens with a leading negative flint, concave on the first surface though weaker than in the Erfle eyepiece. Scaled to a 25 mm focal length, it has broadly similar image quality to the Erfle in Figure 5.12, with astigmatism at high field angles being the most obvious aberration. As with the Erfle eyepiece, the Petzval

(a)
Kidger
Plössl

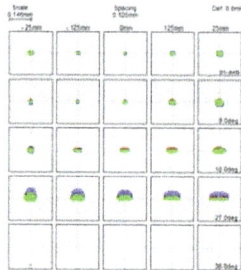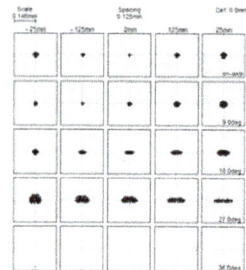

(b)
Erfle
1925
#2

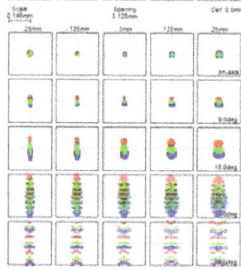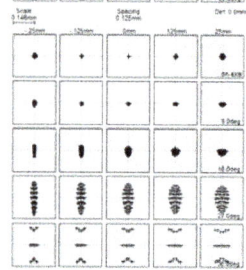

(c)
König
1940
#3

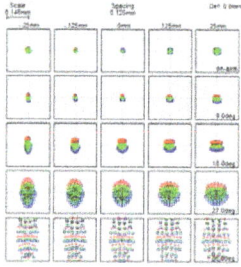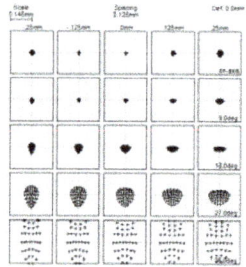

(d)
Nagler
1981
#1

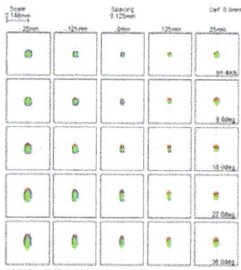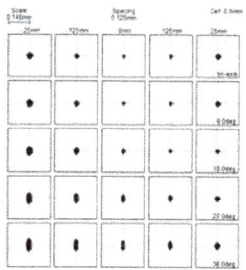

FIGURE 5.13 Wide-field spot diagrams at *f*/10. (a) Kidger's Plössl. (b) Erfle. (c) König. (d) Nagler (Section 5.7) eyepieces. All for focal lengths of 25 mm and field angles of 0°, 9°, 18°, 27° and 36°. Note the scale of the spot diagrams is compressed by a factor of 2 compared to diagrams shown earlier in this book.

curvature places the sagittal focal surface close to the paraxial focus, so off-axis images have a narrow, radially elongated shape. The eye relief is a respectable 22 mm. Design #2 has similar astigmatism but bends the Petzval surface so the paraxial image plane sits roughly midway between the sagittal and tangential astigmatic surfaces, giving the images a more rounded shape. The eye relief in this eyepiece is reduced to 15 mm, but the lateral colour correction is considerably improved. A similar approach to the management of astigmatism is taken in design #3, but the presence of two compound components improves the colour correction (C_{II} = −0.0031 mm for a 24°field) and the eye relief is relaxed slightly to 18 mm.

Robert Tackaberry and Robert Muller patented a similar style eyepiece to König's 1940 design #1 in 1953, having a doublet field lens, single middle lens and single eye lens.[71]

Maximillian Ludewig[72] published five designs, of which numbers 3–5 have an air-spaced doublet field lens but which otherwise are Erfles. Designs 1–2 of that patent are essentially two-group eyepieces in which the third lens is cemented to the fourth as part of an air spaced or cemented triplet eye lens. One potential disadvantage of the designs with more air-spaced elements is an increase in the number of air-glass surfaces, from four in the orthoscopic and Kellners, to six in the Erfle, and eight in Ludewig's variants. Air-glass surfaces have the potential to increase unwanted reflections and ghosts, which diminish the contrast, requiring good antireflection coatings on the surfaces (Section 6.1).

In all five of Ludewig's designs, as in the original Erfle and in Miles' variant, the leading surface of the field lens is concave, on a negative first element which is invariably a flint. It is worth reflecting how these eyepieces, Rank's reversed Kellner and the Plössls bearing meniscus rather than plano-convex components have departed completely from the original field lens concept (Sections 3.4 and 4.1), which served the purpose of directing the principal rays of off-axis objects back towards the optical axis. Recall that this improved the aberrations at the eye lens and reduced the size of the lenses, but also considerably shortened the eye relief. In contrast, leading concave surfaces and leading negative elements now actively divert the principal rays away from the optical axis, as a means of recovering eye relief as well as contributing to the balancing of off-axis aberrations at wider field angles than in earlier eyepieces. We will see this choice used in most of the remaining eyepieces described in this book.

Returning again to the origin of the Erfle as an evolution of the Plössl in which the positive power is shared with an added middle lens, one could go a step further and share the power across *two* middle components, so the eyepiece comprises four weakened components, with the field and eye lenses still being doublets, and the middle group comprising two single-element components. This approach was taken by Wright Scidmore

(1968)[73] and Al Nagler (1985),[74] with Nagler's version (Figure 5.14) distinguished by the field lens and eye lens doublets being not just outwardly concave but having overall negative power. Nagler's eyepiece is impressive, producing good images out to around 30° (24° shown in Figure 5.14) and in fast pencils at $f/6$, and it achieves a 0.02 mm pupil spherical aberration.

The benefit delivered by a negative field element or component may also be understood by comparison with so-called inverted telephoto or retrofocus photographic lenses.[75] In photographic lenses of very short focal length, the desired focal length may be less than the physical spacing between the last lens surface and the focal plane. This can be necessary, for example, in an analogue single-lens reflex camera where a certain amount of space must be reserved behind the lens for the shutter and flip mirror. Short focal lengths can be achieved by using a weak negative first component to "push" the second principal plane of the system rearward of the stronger positive final component. This is similar – but reversed – to the way the trailing negative component of a telephoto lens pushes the second principal plane forward of the first positive component, and how in a Cassegrain telescope the negative convex secondary mirror pushes the second principal plane forward of the positive concave primary mirror.

A later design, which also evolved from an Erfle form and shares a layout outwardly similar to Nagler's 1985 (Figure 5.14a) eyepiece but which looks slightly more like a 4-group, 6-element design, is that by Mitsuhiro Yanari[76] for Nikon®. It has preferably a leading-concave, negative flint element on a positive or meniscus doublet field lens, followed by three further positive components of which one is a doublet. This is distinguished from Nagler's design in which the field lens and eye lens were negative. Yanari's eyepiece targets an eye relief exceeding 80% of the focal length, and a field of view of 60°–70°. Other four-group designs are explored in Section 5.7.

Commercial Eyepieces

With designs continuing to evolve, and brands proliferating and sometimes seeking to distinguish themselves from others without providing meaningful technical specifications, it has become increasingly difficult for the buyer to know quite what they are being offered by some sellers. This can create more confusion than enlightenment, so the following list of contemporary commercial wide-field eyepieces should be read merely as a listing, not as an endorsement or confirmation of any particular design choice or implementation. In some cases, the marketing claims are peppered with apparent inconsistencies which I have not been able to resolve;

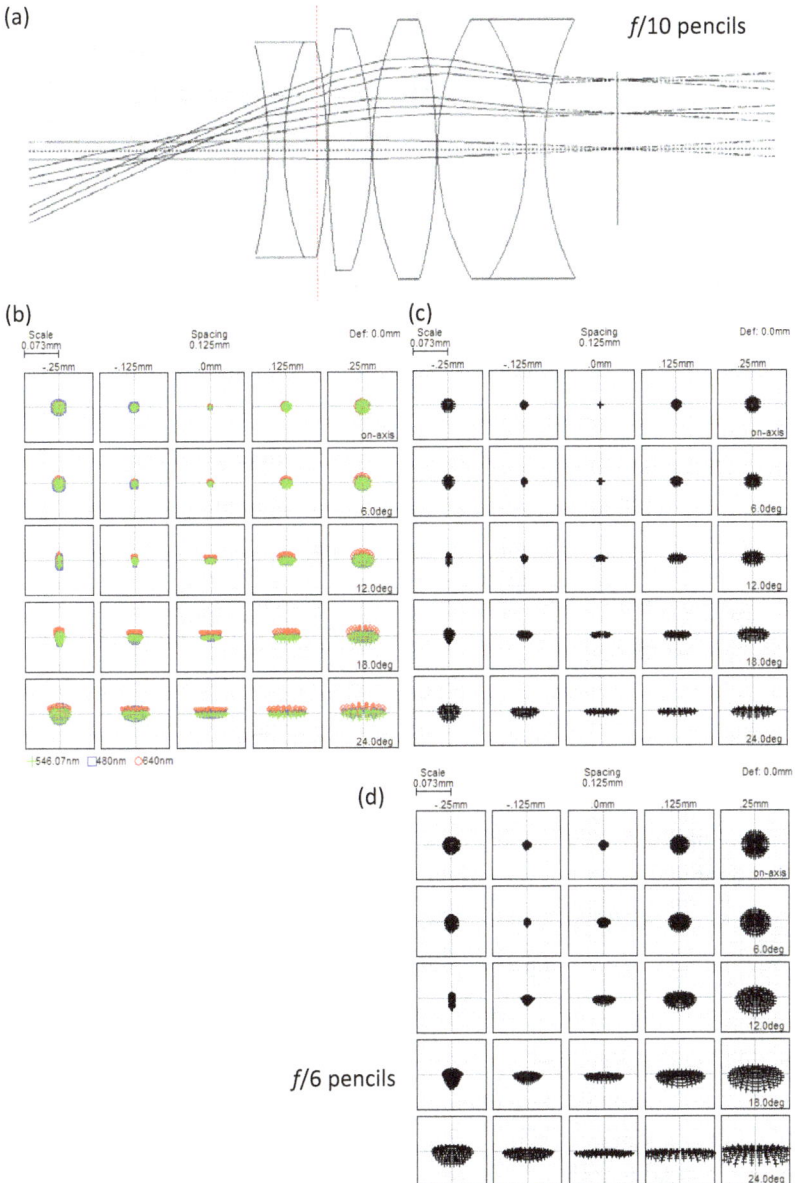

FIGURE 5.14 Four-component Erfle eyepiece scaled from A. Nagler 1985 design [scaled to 25 mm focal length from 40 mm original]. (a) Ray trace. The eyepiece is designed to convey field angles up to 35°. The principal plane is shown in red. (b) Spot diagrams for three wavelengths at apparent field angles of 0°, 6°, 12°, 18° and 24°. (c) Spot diagrams for *f*/10 pencils at apparent field angles of 0°, 6°, 12°, 18° and 24°. (d) Spot diagrams for *f*/6 pencils at apparent field angles of 0°, 6°, 12°, 18° and 24°.

a sceptical reader might even suspect some marketing departments prefer it that way.

Antares® advertises[77] a "W70" series of eyepieces providing a 65° apparent field of view in a 1¼ inch barrel (4–25 mm focal length), and 71°–72° in a 2 inch barrel (31–32 mm focal length), with better eye relief than their Plössl series, but other design details are slim. For some eyepieces, their product notes indicate the number of elements and/or provide some comments on aberrations, suitable *f*/ratios, and/or uses to which the eyepieces might be put.

Celestron's X-Cel LX series (2.5–25 mm focal length) are 6-element, 4-group Erfle-like eyepieces providing a 60° apparent field of view and 16–36 mm eye relief,[78,79] weighing 170–198 g.

Masuyama[80] makes a "Series 53°" (5–30 mm focal length in a 1¼ inch barrel, 35–60 mm in a 2 inch barrel, 53° apparent field of view though 46° in the 60 mm) described as a "Masuyama Ortho Plössl – MOP". As orthoscopic and Plössl eyepieces are different, and classically of a 4-element, 2-group design, the 5-element, 3-group MOP designation clearly suggests an evolved design with at least some features similar to the 3-group, wide-field eyepieces of this chapter.

Omegon has a range of 5- to 6-element "Flatfield" eyepieces (8–27 mm focal length, 9 mm eye relief) giving 53°–60° and up to 4- to 6-element "Premium Flatfield" eyepieces (5.5–25 mm focal length) giving 60°–65° apparent fields of view with 16 mm eye relief. Their "UWA" (ultrawide-angle) eyepieces[81] (6–20 mm focal length) are a 4- to 6-element design, providing 66° apparent field of view and 13–18 mm eye relief. The "SWA" range (10–38 mm) is a 5-element, 3- to 4-group design, providing 70° fields, with a wide weight range from 59 to 693 g.

Sky-Watcher® markets "LE"[82] (1¼ inch barrel, 2–25 mm focal length) and "LET"[83] (2 inch barrel, 28–42 mm focal length) eyepieces with a 20 mm eye relief and fields of view around 45°–56°. The optical design they have adopted is not disclosed. Their "Planetary UWA" series, in contrast, prioritises short focal lengths (2.5–9 mm) suitable for high-magnification planetary observation, providing a 58° apparent field of view in a 5-element, "modified Plössl" form giving a 16 mm eye relief. The Sky-Watcher® Wide/UltraWide series[84] of 6–20 mm focal length, 66° eyepieces likewise has an unspecified design, but shares some characteristics with the Omegon 4-to 6-element UWA series (see above).

Explore Scientific® markets a waterproof 52° Series[85] which employs a 6-element, 3-group design in a 3–20 mm focal-length range, giving

12–17 mm eye relief and weighing 181–227 g, but reverts to a 4-element, 2-group design in the 25–40 mm focal-length range, giving 16–27 mm eye relief at 181–386 g. (The 40 mm eyepiece is in a 2 inch barrel, while the shorter-focal-length eyepieces are 1¼ inch.) It is not clear what optical design has been adopted for Explore Scientific's 62° Series[86] which spans similar focal lengths (5.5–40 mm), eye relief (10–28 mm) and weights (99–323 g where disclosed); note the 32 and 40 mm focal lengths are packaged in 2 inch barrels and weigh 420 and 669 g respectively.

Around the end of the 20th century, Takahashi Seisakusho Ltd produced a range of long eye relief (Takahashi LE) eyepieces using five elements in three groups,[87] intended for use at $f/7$–$f/10$, and providing an apparent field of view of 52°. The 24 mm focal length eyepiece delivered a 17 mm eye relief, which was similar to the Erfles above, though the shorter focal length eyepieces provided a longer eye relief by proportion, e.g., the 7.5 mm focal length LE provided a 10 mm eye relief. The LE line has recently been discontinued, though Takahashi has continued other eyepiece lines.

The Baader planetarium 35 mm "Eudiascopic ED" eyepiece is another 5-element, 3-group eyepiece to have entered and later ceased production.[88] It has been associated[89] with the Erfle class, a modified Plössl ("Super Plössl") design, and a new design in its own right … quite where the design description ends and marketing claims take over is unclear in the absence of technical details. This situation emphasises again the value to purchasers of having clarity over the specification of any eyepieces they are contemplating buying, as for example is provided by the Edmund Optics® website for their RKE® line of eyepieces.

The Vixen® SLV series[90] covers a focal-length range (2.5–25 mm) and offers a 50° apparent field of view, but delivers 20 mm eye relief, and weighs in at 151–176 g; the optical design is not disclosed other than that it employs lanthanum glasses.

5.7 FOUR-GROUP, WIDE-FIELD EYEPIECES

We met the work of Al Nagler in Section 5.5 with a design of modern Plössl eyepieces.[91] He gained hero status amongst amateur astronomers in the 1980s for his designs of advanced wide-field eyepieces which now bear his name. Nagler developed a series of designs, progressively improving upon earlier work, and founded a company – Tele Vue Optics Inc. – which commercialised his eyepieces.[92] His name is now synonymous with eyepieces of this nature, especially those bearing a negative lens upstream of the telescope focal plane. This particular lens has its origins as a standalone accessory called a Barlow lens, so to understand the design of the Nagler eyepieces, it is helpful to examine the Barlow lens first.

Barlow Lens

A Barlow lens, dating from 1834, is a negative power lens that may be inserted into the optical path of a telescope slightly upstream of the focal plane. In reducing the convergence of the pencils coming from the main optics, it increases the effective focal length of the telescope, typically by a factor of around 1.5 or 2 known as the Barlow factor. As the telescope aperture is unchanged, the f/ratio also increases by the Barlow factor. The telescope objective and Barlow lens pair operate in a similar way to a telephoto lens combination, but with the Barlow lens located just a short distance upstream of the objective focal point, whereas in a telephoto lens, the negative component is placed at an intermediate distance from the first lens.[93,94] (See also the comment in Section 5.6 regarding the role of negative secondary optic (mirror) in a Cassegrain telescope.) The Barlow lens is sometimes called a telephoto adapter.

The Barlow lens is sometimes alternatively described as a lens that decreases the effective focal length of the eyepiece. In so far as the angular magnification of a telescope+eyepiece combination is given by the effective focal length ratio $m_{ang} = f'_{tel} / f'_{EP}$, it is not unreasonable to consider the Barlow lens as decreasing the denominator by a factor B rather than increasing the numerator by a factor B as described in the preceding paragraph; it depends whether you consider the Barlow as part of the telescope into which eyepieces are introduced, or whether you consider it as part of the eyepiece.

If we have a telescope of effective focal length f'_1 without a Barlow lens and seek to increase the focal length by a Barlow factor B, then this will deliver a final effective focal length $f'_E = Bf'_1$. Working in terms of powers, we can re-write the separated thin lens equation (Equation 2.17) as $(1/B)F_1 = F_1 + F_B - dF_1F_B$ where F_B is the currently unspecified power of the Barlow lens, d is the separation of the main telescope optic and the Barlow lens, and we recall that $f'_1 = 1/F_1$. The power equation can be rearranged to express the required Barlow lens power as $F_B = -\dfrac{B-1}{B}\dfrac{1}{f'_1-d}$. We note that $f'_1 - d$ is the upstream distance of the Barlow lens from the original telescope focal plane, which we will call v (upsilon) for convenience. We therefore have the requirement that $F_B = -\dfrac{B-1}{Bv}$. Helpfully, this is independent of the telescope aperture or focal length, so Barlow lenses can be designed as a standalone accessory rather than having to be tailored for each telescope. A Barlow factor of 2 could therefore be achieved for an upstream displacement of 4 cm, i.e. 0.04 m, by adopting a Barlow lens of power $F_B = -12.5$ D.

Nagler Eyepieces

While standalone Barlow lenses would usually be achromatised to avoid aberrating the independently designed telescope optics, eyepiece designs that build in a Barlow-like lens have more freedom as to how the aberrations are addressed. The ability of a negative lens placed slightly upstream of the telescope focal plane to address Petzval curvature while retaining the overall positive form of an eyepiece was detailed by Alfred Taylor and Harold Taylor in 1921[95] as an extension to Harold Taylor's previous work[96] on flattening photographic objectives. Their "negative corrector lens" was a four-element (negative, negative, positive, negative) lens situated ahead of three positive groups comprising a concave-first meniscus field lens, a triplet middle lens and a doublet eye lens. Taylor and Taylor's eyepiece was developed for low-power (7× and 10×), small-format telescopes (<25 mm diameter), and employed 6–7 different glass types for the nine elements. Another early design was that by Aaron Levin (1952),[97] using what *might* be considered four groups: two upstream of the telescope focal plane (a negative doublet followed by a positive single lens), and a Ramsden pair downstream. Levin's patent *actually* presented the four groups as a three-group eyepiece (positive single plus Ramsden pair) while the upstream negative group was counted as a distant member of an "objective group" and no prescription was given for it, but the spacings are such that this upstream negative group was located just millimetres from the single positive lens of the "eyepiece group", and a long way from the achromatic doublet primary optic.

    H. Köhler reported a design for a 26.1 mm focal-length eyepiece in 1960, which sought to achieve an apparent field of view of 110°, employed in a 15 × 75 wide-field telescope (binocular).[98] This design likewise made use of a pre-focal-plane Barlow lens comprising three elements, named therein as a Smyth lens. Wright Scidmore and Robert Wolfe patented an eyepiece design in 1969[99] that employed a cemented doublet negative lens upstream of the telescope focal plane, intentionally to increase the eye relief and beneficially to flatten the field of a wide-field eyepiece. The remaining three components of the eyepiece comprised a leading concave surface on a negative-flint-first positive doublet followed by two single positive lenses, similar to a 1940 König eyepiece. Scidmore and Wolfe's work, and Köhler's, was particularly motivated by military sighting applications benefiting from a wide field and large eye relief. Scidmore and Wolfe did not name either the Barlow lens or the Smyth lens as a motivator in their design but nevertheless recognised the role that the negative doublet forward of the focal plane would play.

With hindsight, these various designs represent missed opportunities to accelerate telescope eyepiece design towards a new class of ultrawide flatfield eyepieces. That breakthrough lay with Al Nagler who in the 1980s captivated amateur astronomers with a progression of designs, beginning in 1981[100] with a −39.7 D Barlow[101] situated ahead of three groups broadly resembling an Erfle, but with all four groups optimised together to suppress the aberrations, with the eyepiece incorporating doublets in at least three components, and with the design employing just 2 glass types. The eyepieces were designed as 10 mm focal lengths, with 17 mm eye relief and an apparent field of view to 90°. (With the Barlow lens eliminated, the focal length of the Erfle-like portion would be 22.60 mm, which echoes the alternative description above of a Barlow as reducing the focal length of the eyepiece.) In its four-group form, Nagler's 10 mm focal-length eyepiece of 1981 delivered excellent image quality. For comparison with the other 25 mm focal-length eyepieces presented in this book, I show in Figure 5.15 a rescaled 25 mm focal-length version of Nagler's 10 mm focal-length 1981 #1 lens layout, with ray paths, spot diagrams and Seidel aberrations for field angles up to 24° at $f/10$, and in Figure 5.15e at $f/6$. The remarkable performance of the eyepiece is evident at first glance, with well-focused images.

The principal reason for the transformational image quality of this eyepiece in comparison to all forerunners can be seen in the Seidel aberration coefficients in Figure 5.15d. In all previous eyepieces, the three biggest challenges have been astigmatism, Petzval curvature and lateral chromatic aberration, with pupil spherical aberration and poor eye relief providing further challenges. The 1981 Nagler dealt comprehensively with the first two of these via the Barlow lens. We noted in Section 3.7 that field curvature depends substantially on the sum of the lens powers but does not depend on the location of the lens in the train. Nagler's inclusion of a strong negative-power Barlow lens close to the telescope focal plane in an otherwise positive eyepiece allows Petzval curvature to be almost nulled. Astigmatism, which we recall also scales with lens power, is all but defeated by the same negative component; while we have typically

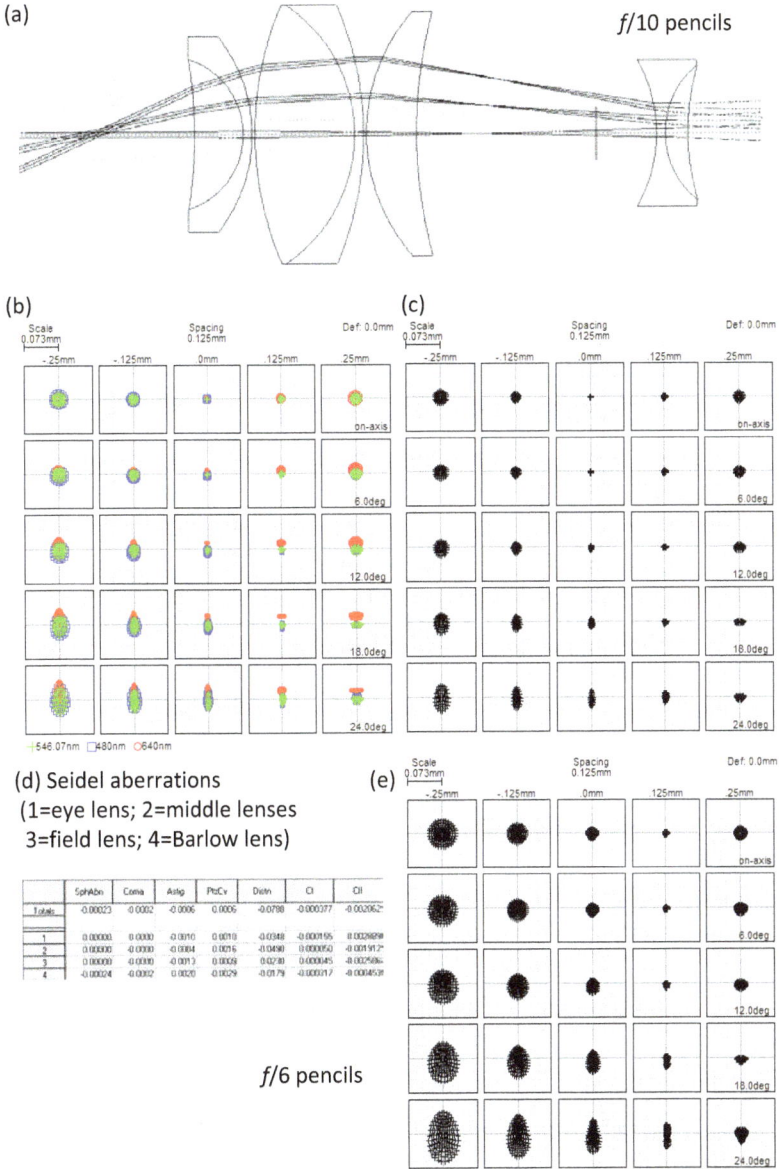

FIGURE 5.15 Nagler 1981 10 mm focal-length eyepiece scaled to 25 mm [25 mm focal length, F_B = −15.9 D, F_{FL} = 5.4 D, F_C = 8.1 D, F_{EL} = 5.5 D]. (a) Ray trace for field angles of 0°, 12° and 24°. (b) Spot diagrams for three wavelengths at apparent field angles of 0°, 6°, 12°, 18° and 24°. (c) Spot diagrams for f/10 pencils at apparent field angles of 0°, 6°, 12°, 18° and 24°. (d) Seidel wavefront aberrations (f/10, 24°) by component. (e) Spot diagrams for f/6 pencils at apparent field angles of 0°, 6°, 12°, 18° and 24°.

seen S_{III} coefficients around −0.0030 or so in the various eyepieces we have examined, Nagler's Barlow introduces +0.0020 (in the scaled-up version of Figure 5.15) which preconditions the pencils for the astigmatism that the later positive components introduce. Finally, by having three doublets within the four components, there is sufficient scope to suppress chromatic aberration.

The 25 mm focal-length eyepiece also achieves a long eye relief of 47 mm, but for all its accomplishments, the scaled-up design inevitably has a few drawbacks. The first, evident from the ray paths in Figure 5.15a, is that the 12° and 24° pencils do not cross the optical axis in the same location, i.e. there is considerable pupil spherical aberration (0.09 mm). The second is the sheer size and weight of the eyepiece, which could make it cumbersome to use on a small telescope and expensive to procure, and explains why Nagler designed to shorter focal-lengths than the 25 mm monster simulated here for comparison purposes. Nevertheless, the design has a further advantage, which is that its excellent imaging quality is maintained out to high field angles; spot diagrams out to 36° are shown in Figure 5.13d which are better corrected than any of the other designs shown in that figure.

The challenges of the scaled-up design in Figure 5.15, noted in the preceding paragraph, limited the commercial implementation to short-focal-length forms, but these limitations (especially the pupil spherical aberration) were explicitly addressed in a subsequent (1987) design,[102] which sought to improve the eyepiece further and make it practical at longer focal lengths. The 1987 eyepiece also increased the number of positive components from three to four. Nagler noted that to reduce pupil spherical aberration, the surface of the first positive element facing the Barlow lens must be convex or plano, not concave as in his 1981 design in which the positive components adopted a more Erfle-like form, where a strong leading concave surface is common. Nagler's 1987 design therefore marks a significant evolution beyond the Barlow+Erfle form into a more distinctive eyepiece, albeit continuing to exploit the design opportunities afforded by his Barlow implementation. One of the eyepieces shown in the design, #4, implements a split form for the Barlow in a further evolution, and produces much improved pupil spherical aberration. In a rescaled simulation with a 25 mm focal length (Figure 5.16), this eyepiece achieves a Seidel pupil spherical aberration of just 0.03 mm for $f/10$ pencils and a 24° field angle (cf. 0.09 for the 1981 eyepiece), and manages a 13 mm eye relief, though it also exhibits poorer chromatic aberration than the 1981 eyepiece and we see coma in the off-axis images.

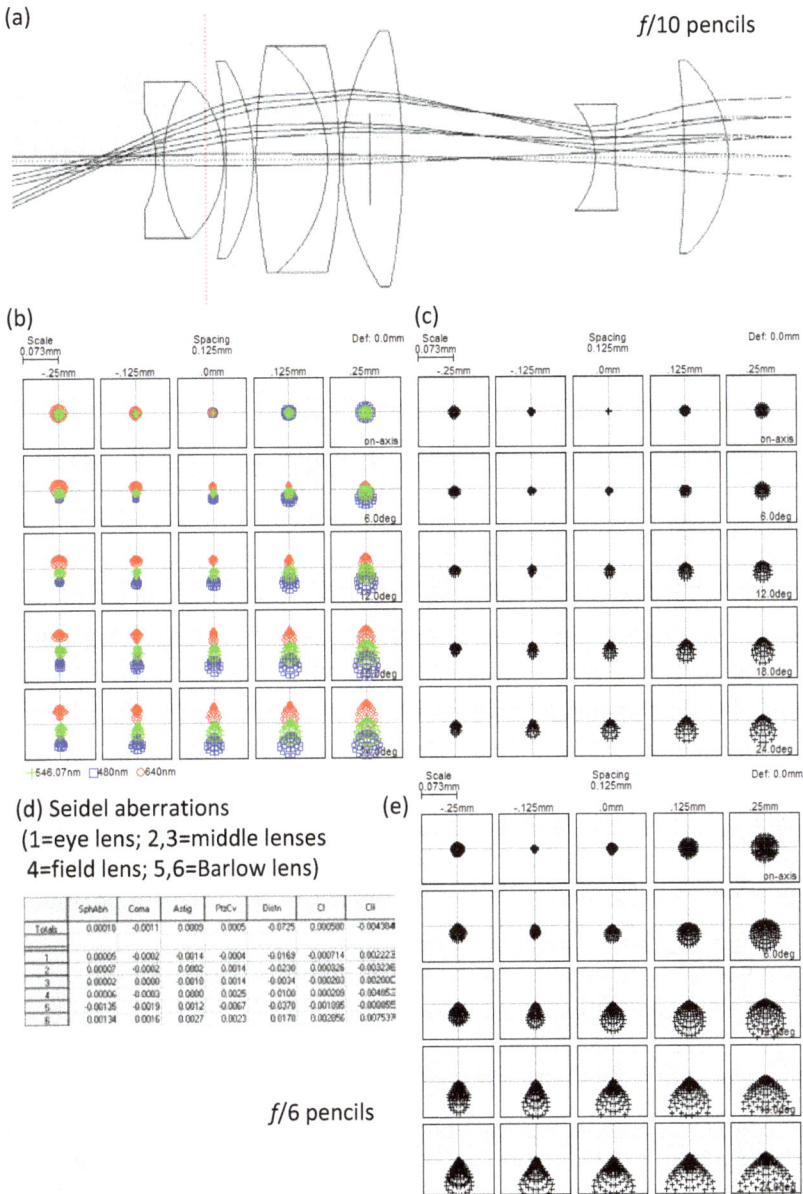

(a) f/10 pencils

(b)

(c)

(d) Seidel aberrations
(1=eye lens; 2,3=middle lenses
4=field lens; 5,6=Barlow lens)

	SphAbn	Coma	Astig	PtzCv	Distn	CI	CII
Totals	0.00010	-0.0011	0.0009	0.0005	-0.0725	0.000500	-0.004384
1	0.00005	-0.0002	-0.0014	-0.0004	-0.0169	-0.000714	0.002223
2	0.00007	-0.0002	0.0002	0.0014	-0.0230	0.000326	-0.003236
3	0.00002	0.0000	0.0010	0.0014	-0.0034	-0.000203	0.002000
4	0.00006	-0.0003	0.0000	0.0025	-0.0100	0.000209	-0.004653
5	-0.00125	-0.0019	0.0012	-0.0067	-0.0370	-0.001895	-0.000955
6	0.00134	0.0016	0.0027	0.0023	0.0178	0.002056	0.007537

(e)

f/6 pencils

FIGURE 5.16 Nagler 1987 10 mm focal-length eyepiece scaled to 25 mm. (a) Ray trace for field angles of 0°, 12° and 24°. The principal plane is shown in red. (b) Spot diagrams for three wavelengths at apparent field angles of 0°, 6°, 12°, 18° and 24°. (c) Spot diagrams for f/10 pencils at apparent field angles of 0°, 6°, 12°, 18° and 24°. (d) Seidel wavefront aberrations (f/10, 24°) by component. (e) Spot diagrams for f/6 pencils at apparent field angles of 0°, 6°, 12°, 18° and 24°.

Donald Dilworth[103] for Optical Systems Design, Saburo Sugawara[104] for Canon, and Yang Xihua[105] for Kunming Ruishi Optical Instrument Manufacturing Co. explored similar solutions involving a negative lens paired with another ahead of the telescope focal plane, and additional groups further downstream. Dilworth targeted a 90° field of view, Sugawara's challenge was to obtain an apparent field of view exceeding 75°, and Yang targeted 110° with an 8-element, 6-component design. Sugawara particularly cited a desire to avoid the eyepiece becoming too cumbersome to use, and also described a more compact design, achieving a 60° field. In a significant departure from previous eyepieces that I have discussed so far, it is notable that four of the eleven designs patented by Sugawara utilised an aspheric surface, an approach we discuss in Section 6.3. A further design, by Ruisi Fu[106] for Kunming Jinghua Optical Co., used 12 elements in 8 groups to target a 120° field, necessarily evolving further from Nagler's design.

While Nagler's 1981 eyepiece began with and extended a Barlow+Erfle model, the choice of an Erfle model for the positive groups is not essential. Thomas Clarke presents[107] a relatively simple three-element eyepiece based on a Barlow+Huygens pairing, where the Barlow strength is chosen to null the Petzval curvature of the Huygens, and where the simplicity of a single glass type (albeit a lanthanum crown) is retained, along with other economies such as symmetric and plano-surfaced lens forms. As noted already, adoption of a lanthanum glass helps reduce both monochromatic and chromatic aberrations. Clarke presents several designs, and acknowledged room for further refinement if the lenses were re-bent or split, and if a different, higher-dispersion glass was introduced for the field lens. His "Figure 5" design, further modified by using BASF52 glass for the field lens instead of LAK10 following his suggestion, yields the encouraging spot diagrams in Figure 5.17. (LAK10 and BASF52 have similar d-line refractive indices $n_d = 1.72$ and 1.70, but different Abbe values, $V_d = 50$ and 44; see also Figure 2.3b.) The lens powers ($F_B = -39$ D, $F_{FL} = +19$ D, $F_{EL} = +19$ D) have a Petzval sum (Equation 3.8) close to zero as intended. The eye relief in this particular design is a staggering 41 mm, though the pupil spherical aberration is 0.16 mm which is significant. The image quality deteriorates rapidly beyond field angles of 20°, so this is not a wide-field eyepiece, but a 40° apparent field of view is still a significant accomplishment, especially as further optimisation is surely possible.

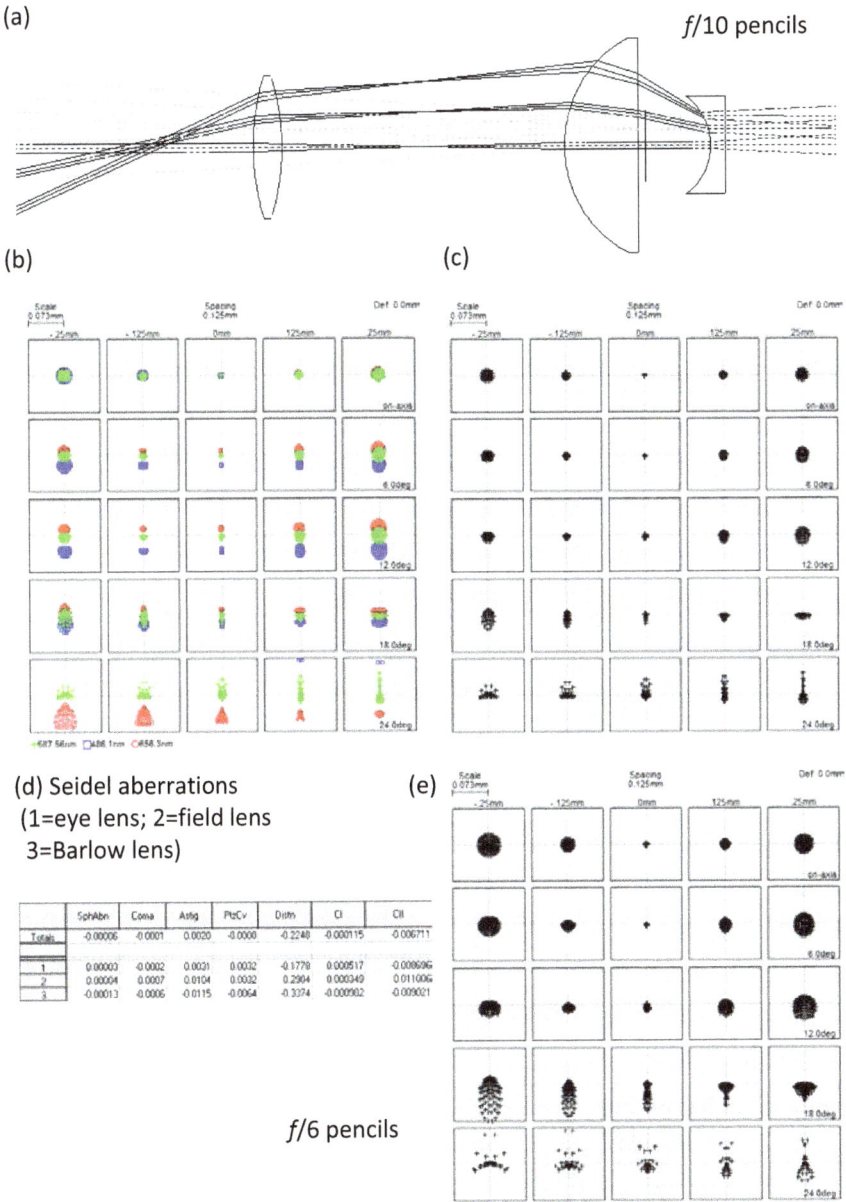

(a) f/10 pencils

(b)

(c)

(d) Seidel aberrations
(1=eye lens; 2=field lens
3=Barlow lens)

(e)

	SphAbn	Coma	Astig	PtzCv	Dstn	CI	CII
Totals	-0.00006	-0.0001	0.0020	-0.0000	-0.2248	-0.000115	-0.006711
1	0.00003	-0.0002	0.0031	0.0032	-0.1778	0.000517	-0.008696
2	0.00004	0.0007	0.0104	0.0032	0.2904	0.000349	0.011006
3	-0.00013	-0.0006	-0.0115	-0.0064	-0.3374	-0.000982	-0.009021

f/6 pencils

FIGURE 5.17   Barlow + Huygens design by T. Clarke (1983) 25 mm focal-length (a) Ray trace for field angles of 0°, 12° and 24°.(b) Spot diagrams for three wavelengths at apparent field angles of 0°, 6°, 12°, 18° and 24°. (c) Spot diagrams for f/10 pencils at apparent field angles of 0°, 6°, 12°, 18° and 24°. (d) Seidel wavefront aberrations (f/10, 24°) by component. (e) Spot diagrams for f/6 pencils at apparent field angles of 0°, 6°, 12°, 18° and 24°.

Other Commercial Eyepieces

Baader planetarium sells an 8-element, 5-group "Hyperion®" series[108] with a 68° apparent field of view. The range spans focal lengths 5–24 mm, eye reliefs from 17 to 22 mm, and they weigh from 311 to 406 g. The Hyperion range is advertised with a "Multiple Focal Length (MFL)" capability[109]; in practice this means the front portion of these assemblies can be disconnected, which temporarily removes the Barlow-like front group of lenses located in the nosepiece of the eyepiece and leaves a longer focal-length eyepiece intact in the main portion of the eyepiece assembly, having a focal length around 22 mm (or 32 mm for the 21 mm focal length Hyperion). This feature was pre-empted above in connection with Nagler's designs, where it was noted that the first, negative Barlow group sits in front of what in that case was an Erfle-like positive assembly; a similar concept was seen in Scidmore and Wolfe's Barlow+König construction.[110] It is interesting to see this capability made core to the Hyperion design, and the flexibility provided to remove and reattach the Barlow group at will.

The Celestron® Ultima Edge series[111] is a 5- to 9-element eyepiece in 4–5 groups, but of otherwise unspecified design, in 10–30 mm focal lengths, with an apparent field of view of 60°–65° (70° in the 30 mm focal length eyepiece which has a 2 inch barrel), and 16–22 mm eye relief.

Explore Scientific® markets a 68° Series[112] spanning 16–40 mm focal lengths.

Meade® produced a "Series 5000 HD-60" eyepiece range[113] (4.5–25 mm focal length, 1¼ inch barrel) comprising 6 elements, yielding a 60° apparent field of view and eye relief of 19 mm. The "Series 5000 UHD" range (10–30 mm focal length) is a 5- to 9-element series with apparent fields of 60°–70°. (See also note 60.)

Omegon's "Cronus WA" (wide angle) series[114] eyepieces have short focal lengths (2.5–9 mm) mostly suited for planetary observing, but use a 6-element, 4-group design to provide a 60° field of view, with 16 mm eye relief, which would not be possible in orthoscopic or Plössl styles at these focal lengths. They weigh in at 136–154 g. Their "Redline SW" series,[115] on the other hand, is an 8-element design attaining a 70° field (3.5–32 mm focal lengths) with 20 mm eye relief. The Omegon "Super LE" series[116] (7–18 mm focal length) has an 8-element, 4-group design that provides a 68° field of view with 20 mm eye relief, having 313–330 g weights, while the "LE Planetary" series[117] (3–18 mm focal length) attain a 55° field using a 7-element, 4-group design (5-element, 3-group for the 18 mm) with 20 mm eye relief, and weights 149–252 g.

Pentax have developed their XW series[118] (3.5–40 mm) encompassing a wide range of designs; they are not simply rescaled versions of one base design. They contain 6- to 8-elements in 4- to 7-groups. Most are Barlow-(or Smyth-)lens-equipped designs, all of varying form, while the 40 mm resembles an Erfle with a triplet eye lens. They provide an eye relief of 20 mm, and an apparent field of view of 70°, though the 16.6 and 23 mm extend to 85°. Weighing from 335 to 770 g for most, and a hefty 4055 g for the 3.5 mm (that's how Pentax achieves a 20 mm eye relief for a 3.5 mm focal length eyepiece), these are not intended for very small telescopes.

Sky-Watcher and Optical Vision Ltd (OVL) have both been associated with a PanaView® 2 inch, 5-element eyepiece (26–38 mm focal length) giving a 70° apparent field of view, and eye relief of 20–28 mm.

The Takahashi "TOE" series[119] comprises very short focal-length eyepieces (2.5–4 mm) intended for high-magnification applications and delivers a 52° apparent field of view. This field is not nearly as large as the Nagler eyepieces listed above but is more than a traditional orthoscopic (though see note 26), and comparable to Plössls. The TOE eyepieces also deliver eye relief of 10 mm, more than twice the focal length. The Takahashi TOE series has 6 elements over 4 groups, and they describe the first of these as a Smyth lens, which is an integrated Barlow by another name.[120] With weights of 140–145 g, these are physically smaller assemblies than the original Nagler eyepieces described above or the ultrawide-field eyepieces we shall meet in Section 5.8.

In addition to ultrawide-field eyepieces we shall meet in Section 5.8, Tele Vue® also markets[121] the "DeLite" series (a smaller, lighter and more economical version of their "Delos") offering a 62° apparent field of view in 3–18.2 mm focal lengths and weighing 200–220 g), and the "Panoptic" (68° field of view in 19–41 mm focal lengths, 187–953 g).

Vixen® produces a 42 mm focal length "LVW" eyepiece[122] with a 65° apparent field of view in a 2 inch barrel, giving a 20 mm eye relief, weighing 543 g.

5.8 ULTRAWIDE-FIELD EYEPIECES

What counts as an ultra wide-field eyepiece? I have adopted a fairly arbitrary threshold of 72° or more apparent field of view; other criteria may be preferred. Barrel sizes of 2 inches are more common amongst the wider-field eyepieces because of the large field stops needed to accommodate such large fields. The possible exceptions are at the shorter focal lengths; a 72° eyepiece with focal length <20 mm could, in principle, utilise a 1¼ inch barrel.

With eyepieces now delivering apparent fields of view up to 110°, it is worth noting that the field of vision of the human eye[123,124] is around 120°, and we have already noted that away from the central foveal region, the resolution of the eye is much reduced due to the lower density of rod cells and the connection of multiple cells into a smaller number of nerve fibres that delocalise the signal transmitted to the brain (Section 3.7). Future developments in eyepieces are unlikely to benefit from increases in the field of view beyond the current maximum.

As foreshadowed above, ultrawide-field eyepieces have incredible optical capabilities compared to eyepieces of 50 years ago, but achieving this requires a lot more glass than historically. For smaller telescopes, the weight of that may be too much. Several currently available eyepieces are summarised in the following subsection. The examples discussed have been chosen because of their relevance to the topic of this section; the list is not intended to be complete and serves neither as an endorsement nor a criticism of particular manufacturers' approaches to eyepiece design, but at least it provides a snapshot of some of the characteristics of ultrawide-field eyepieces available in 2025. In recognition of Al Nagler's trailblazing role in developing modern ultrawide-field eyepieces, I begin with the Tele Vue® eyepieces, reverting to alphabetical order after that.

Tele Vue®

The 1981–1987 Nagler designs (Section 5.7) ushered in a new generation of ultrawide-field eyepieces.[125] As of 2025,[126] Tele Vue® had replaced it earliest Nagler⁰ eyepiece designs (original and Type-2) with the long-eye-relief 7-element, 5-group Type-4 (22 mm focal length, weighing 680 g), 6-element, 4-group Type-5 (16 and 31 mm focal length, 201 g and 998 g respectively), the 7-element, 4-group compact Type-6 (3.5–13 mm focal length, 181–241 g) and long-eye-relief Type-7 (5.5–19 mm focal length, 550–584 g) eyepieces, all delivering an 82° apparent field of view. The company was also offering the "Ethos" (100°–110° apparent field of view in 3.7–21 mm focal lengths, weighing from 430 to 1020 g) launched in 2007, and the "Delos" (72° apparent field of view in 3.5–17.3 mm focal lengths weighing 408–499 g).

A-Z

The Antares® Speers-Waler Series-3 and 4 eyepieces[127] (focal lengths 6–25 mm) offer 74–86° apparent fields of view, with typically 6–8 elements. They also market a 100°, 20 mm focal length "XWA" eyepiece.

Baader Planetarium sells an 8-element, 5-group "Morpheus® 72°" series[128] with a 72° apparent field of view. The Morpheus range spans focal lengths from 4.5 to 17.5 mm, eye relief from 17 to 23 mm, and weighing from 345 to 415 g.

Celestron® has developed its Luminos® series[129] (7–23 mm focal length) as 6- to 8-element, 4- to 5-group eyepieces providing an 82° apparent field of view and 12–27 mm eye relief. The specific design is not disclosed, but their weights span ~340–900 g.

Explore Scientific has a number of ultrawide-angle series[130,131,132,133] providing 82°, 92°, 100° and 120° apparent fields of view. The weights of some of the eyepieces are considerable, in view of the amount of glass required to deliver imaging over such angles.

The Masuyama 85° Series[134] (10–32 mm focal length) is a 5-element ultrawide-field eyepiece.

The Meade® 5000 series includes 7-element "UWA" (ultrawide angle, 82°, 5.5–20 mm focal length), "PWA" (premium wide angle, 82°, 4–28 mm focal length) and 7- to 9-element "MWA" (mega wide angle, 100°, 5–21 mm focal length) variants.

Omegon provide the 8-element, 4-group "OGDO" series[135] (4–20 mm focal length) of 80° eyepieces with 20 mm eye relief, using lanthanum glass, the "Oberon" series[136] (7–32 mm focal length) of 82° eyepieces offering 12–27 mm eye relief, and the 7- to 9-element "Panorama II"[137] (5–21 mm) at 100°, giving 13–20 mm eye relief. All three series are waterproof and weigh in the range 245–947 g.

The Pentax XW16.5 and XW23 (see Section 5.7) attain fields of view of 85°, and thus fall into this ultrawide-field category.

The Sky-Watcher® "Panorama" series[138] (7–23 mm focal length) is a 7-element eyepiece delivering an 82° apparent field of view, with weights from 350 to 830 g. The specific design is not disclosed. The "Nirvana™-ES UWA" (ultrawide angle) series[139] is a 7-element, 4-group 82° eyepiece, in 4–16 mm focal lengths intended for planetary observation, and providing 12 mm eye relief. Weighting in close to 170 g each, they are a physically smaller eyepiece than many of the other ultrawide-field eyepieces. The series has been attributed sometimes to Sky-Watcher, and sometimes to OVL (Optical Vision Limited), which is a distributor for Sky-Watcher amongst other optical manufacturers.

Most of the analysis in this book has been of eyepieces having a focal length of 25 mm on telescopes feeding either $f/10$ or $f/6$ pencils, in order that the eyepieces' features could be appreciated without too many variable factors coming into play. Obviously, people observe using a range of eyepiece focal lengths in order to vary the magnification and field of view presented to their eye. Some properties of eyepieces, such as eye relief and true field of view, scale reasonably linearly with focal length within a given design, but upon switching design different factors come into play. As you contemplate using eyepieces of other focal lengths, bear in mind that their performance may vary from that shown in the illustrations in this book, and as emphasised in Section 5.1, there is considerable diversity even within a given eyepiece type.

The discussion so far has concentrated solely on spherical optics. Both the optics theory introduced in Chapter 2, and the eyepiece investigations in Chapters 3–5 assumed that all refracting surfaces would be spherical, since these are the easiest to describe geometrically and therefore the easiest to model mathematically and computationally, and also the easiest to manufacture and test. However, we saw in Chapters 3–5 that spherical surfaces are far from perfect at bringing the rays from a given pencil to a common focus, so we need to also consider the possibility that eyepieces could be improved if non-spherical surfaces were adopted. We consider aspherics (non-spherical surfaces) in Chapter 6, alongside other topics important to eyepieces that go beyond a description of the glass lenses from which they are assembled.

NOTES

1 R. Kingslake and R.B. Johnson, *Lens Design Fundamentals*, 2nd edition, Academic Press, Chapter 7.

2 Carl Zeiss (co.), Ramsdensches Okular mit einem zusammengesetzten Augenlinsensystem, in dem eine chromatisch korrigierende Kittfläche ihre konkave Seite der Feldlinse zukehrt, Kaiser Königl. Patentamt, Austrian Patent AT29612, 1907 https://worldwide.espacenet.com/patent/search/family/003543116/publication/AT29612B?q=pn%3DAT29612B (accessed 20/02/2025).

3 M. Kidger, *Fundamental Optical Design*, SPIE, 2000, Chapter 6.

4 A. König, Okular, Deutsches Patentamt, Patent DE898084, 1953 https://worldwide.espacenet.com/patent/search/family/007617816/publication/DE898084C?q=pn%3DDE898084C (accessed 16/02/2025).

5 M.J. Kidger, *Fundamental Optical Design*, SPIE, 2000, Chapter 11.

6 The air in the Earth's atmosphere has a refractive index slightly above 1.0, typically 1.0003 at ground level, which decreases towards 1.0 as the air density falls at greater height. As rays from astronomical objects pass some tens of kilometres through the Earth's atmosphere, they experience sufficient dispersion to noticeably split the red and blue light making up the ground-based image of the source, particularly for objects with a

small angular extent such as stars and planets. Atmospheric dispersion is greater for objects at lower altitude angles, i.e. nearer the horizon, due to the greater atmospheric path length traversed on those lines of sight. Objects observed near the horizon, particularly at high magnification, are likely to exhibit degraded images having coloured fringes because of atmospheric dispersion and are better observed when they attain higher altitude angles. Observing instruments on some professional telescopes are equipped with atmospheric dispersion compensators to correct for this effect.

7 Distortion may be assessed in different ways, depending on the application. As an eyepiece is intended to convey a more-or-less linear (flat) object in the telescope focal plane as a visually observed virtual image at infinity, ideally equally spaced linear distances h in the focal plane will be rendered as equal intervals of angular separation θ' in the image space, with $h = f'\theta'$. See M. Kidger, *Fundamental Optical Design*, SPIE, 2000, Chapter 4, Section 4.4.2.4, and H. Rutten and M. van Venrooij, *Telescope Optics Evaluation and Design*, Willmann-Bell Inc., Chapter 16.3. We therefore show the "f-theta" form of distortion.

8 M.J. Kidger, *Fundamental Optical Design*, SPIE, 2000, Chapter 4.

9 "Generally" in this case refers to the most commonly illustrated form of spherical aberration, where marginal rays cross the principal ray ahead of the paraxial rays. This is called positive spherical aberration, and is the common situation for single, positive-powered lenses. Spherical aberration can, as we have seen, be varied but not eliminated by bending a single lens. As eyepieces are net-positive systems, spherical aberration is generally of the positive variety. Optical systems with negative spherical aberration certainly exist but will be ignored in our focus on eyepieces.

10 M.J. Kidger, *Fundamental Optical Design*, SPIE, 2000, Chapter 4.

11 M.J. Kidger, *Fundamental Optical Design*, SPIE, 2000, Chapter 4.

12 M.J. Kidger, *Fundamental Optical Design*, SPIE, 2000, Chapter 4.

13 M.J. Kidger, *Fundamental Optical Design*, SPIE, 2000, Chapter 4.

14 M.J. Kidger, *Fundamental Optical Design*, SPIE, 2000, Chapter 10.

15 M.J. Kidger, *Fundamental Optical Design*, SPIE, 2000, Chapter 4.

16 M. von Rohr, Improvement in Magnifying Lenses, Eye-pieces and the like, United Kingdom Patent GB24009, 1903 https://worldwide.espacenet.com/patent/search/family/032160518/publication/GB190324009A?q=pn%3D190324009 (accessed 20/02/2025). A. König, Ramsden Ocular, US Patent US873871, 1907 https://worldwide.espacenet.com/patent/search/family/002942315/publication/US873871A?q=pn%3DUS873871 (accessed 20/02/2025). Carl Zeiss (co), Oculaire de ramsden à verre d'oeil composé d'une lentille collectrice simple et d'une lentille double, divergente ou convergente, Office National De La Propriété Industrielle, French Patent FR375748, 1907 https://worldwide.espacenet.com/patent/search/family/001422289/publication/FR375748A?q=pn%3DFR375748 (accessed 20/02/2025).

17 H. Erfle, Eyelens system, United States Patent Office, US1479229, 1924 https://worldwide.espacenet.com/patent/search/family/023954988/publication/US1479229A?q=pn%3Dus1479229 (accessed 26/02/2025).

18 A. Nagler, Plössl Type Eyepiece for Use in Astronomical Instruments, US Patent 4482217, 1984, https://worldwide.espacenet.com/patent/search/family/023857522/publication/US4482217A?q=pn%3Dus4482217 (accessed 23/02/2025).

19 Celestron® Eyepiece Specifications, 2023 https://www.celestron.com/blogs/knowledgebase/how-do-i-use-my-telescope-eyepieces (accessed 07/03/2025).

20 Edmund Optics RKE® Precision Eyepieces, https://www.edmundoptics.co.uk/f/edmund-optics-rke-precision-eyepieces/12484/ (accessed 27/02/2025). Edmund Optics® provide detailed optical design information in downloadable text files (.zmx extension, but just a text file) compatible with the optical design software commonly known as Zemax.

21 M. Kanai, Eyepiece System, US Patent, US5790313A, 1995.

22 L. Bertele, Okular, Reichtspatentamt, German Patent DE570499, 1933 https://worldwide.espacenet.com/patent/search/family/006568601/publication/DE570499C?q=pn%3DDE570499C (accessed 20/02/2025).

23 Edmund Optics RKE® Precision Eyepieces, https://www.edmundoptics.co.uk/f/edmund-optics-rke-precision-eyepieces/12484/ (accessed 27/02/2025). Edmund Optics® provide detailed optical design information in downloadable text files (.zmx extension, but just a text file) compatible with the optical design software commonly known as Zemax.

24 H.E. Bennett, J.E. Bennett and A.H. Guenther, In Memoriam David Herr Rank, *Journal of the Optical Society of America A*, 71(9), 1148–1149, 1981.

25 Carl Zeiss (co.), Improvements in Telescope Eye-pieces, United Kingdom Patent, GB509585, 1939 https://worldwide.espacenet.com/patent/search/family/005794114/publication/GB509585A?q=pn%3DGB509585 (accessed 19/02/2025).

26 Baader Classic Ortho/Plössl (Q-Turret) 1¼″ Eyepieces https://www.baader-planetarium.com/en/classic-ortho-10mm-1-25inch-eyepiece-ht-mc.html (accessed 30/07/2025).

27 L. Bertele, Okular, Reichtspatentamt, German Patent DE570499, 1933 https://worldwide.espacenet.com/patent/search/family/006568601/publication/DE570499C?q=pn%3DDE570499C (accessed 20/02/2025).

28 Baader Classic Ortho/Plössl (Q-Turret) 1¼″ Eyepieces https://www.baader-planetarium.com/en/classic-ortho-10mm-1-25inch-eyepiece-ht-mc.html (accessed 30/07/2025).

29 Masuyama is the eyepiece brand manufactured by Ohi Kohki Co., Ltd, Japan, named for the optical designer Ichiharu Masuyama who founded the company in the mid-20th century. Source: J.R. Dire, The Masuyama 1.25″ 53° Premium Eyepieces https://www.hutech.com/file/Masuyama%20Eyepieces.pdf (accessed 08/03/2025).

30 Omegon orthoscopic eyepieces https://www.omegon.eu/telescope-accessories/eyepieces/15_10/m,Omegon/a,Zubehoer.Allgemein.Serie=Ortho (accessed 08/03/2025).

31 Takahashi Abbe eyepieces https://takahashi-europe.com/catalog/accessories/eyepieces/abbe-eyepieces/abbe-eyepieces (accessed 08/03/2025).

32 A. König, Okular, Deutsches Patentamt, Patent DE898084, 1953 https://worldwide.espacenet.com/patent/search/family/007617816/publication/DE898084C?q=pn%3DDE898084C (accessed 16/02/2025).

33 A. König, Ramsden Ocular, US Patent US873871, 1907 https://worldwide.espacenet.com/patent/search/family/002942315/publication/US873871A?q=pn%3DUS873871 (accessed 20/02/2025).

34 A. König, Ocular, United States Patent Office US1159223, 1915 https://worldwide.espacenet.com/patent/search/family/003227279/publication/US1159233A?q=pn%3DUS1159233A (accessed 24/04/2025).

35 H. Rutten and M. van Venrooij, *Telescope Optics Evaluation and Design*, Willmann-Bell Inc., 1988, Chapter 16.

36 The König eyepiece was in particular marketed by a company called University Optics, which supplied eyepieces and other astronomical instruments and accessories from around 1962 to 2017.

37 H. Rutten and M. van Venrooij, *Telescope Optics Evaluation and Design*, Willmann-Bell Inc., 1988, Chapter 16.

38 I. Ridpath (ed.), *Norton's Star Atlas and Reference Handbook*, 20th edition, Pi Press, 2004, Chapter 2.

39 R. Kingslake and R.B. Johnson, *Lens Design Fundamentals*, 2nd edition, Academic Press, 2010, Chapter 16.

40 A. Taylor and H.D. Taylor, Improvements in Eyepieces for Telescopes and Other Optical Instruments, United Kingdom Patent, GB126837, 1919 https://worldwide.espacenet.com/patent/search/family/032343803/publication/GB126837A?q=pn%3DGB126837 (accessed 21/02/2025).

41 M.J. Kidger, *Fundamental Optical Design*, SPIE, 2000, Chapter 10.

42 R. Kingslake and R.B. Johnson, *Lens Design Fundamentals*, 2nd edition, Academic Press, 2010, Chapter 12.

43 R. Kingslake and R.B. Johnson, *Lens Design Fundamentals*, 2nd edition, Academic Press, 2010, Chapter 13.

44 M.J. Kidger, *Fundamental Optical Design*, SPIE, 2000, Chapter 7.

45 M.J. Kidger, *Fundamental Optical Design*, SPIE, 2000, Chapter 7.

46 Carl Zeiss (co.), Improvements in Telescope Eye-pieces, United Kingdom Patent, GB509585, 1939 https://worldwide.espacenet.com/patent/search/family/005794114/publication/GB509585A?q=pn%3DGB509585 (accessed 19/02/2025).

47 A. Nagler, Plössl Type Eyepiece for Use in Astronomical Instruments, US Patent 4482217, 1984, https://worldwide.espacenet.com/patent/search/family/023857522/publication/US4482217A?q=pn%3DUs4482217 (accessed 23/02/2025).

48 M.J. Kidger, *Fundamental Optical Design*, SPIE, 2000, Chapter 11.

49 Antares® eyepieces https://antares.space/eyepieces (accessed 07/03/2025).

50 Baader Classic Ortho/Plössl (Q-Turret) 1¼″ Eyepieces https://www.baader-planetarium.com/en/classic-ortho-10mm-1-25inch-eyepiece-ht-mc.html (accessed 30/07/2025).

51 Bresser® eyepieces https://www.bresseruk.com/Astronomy/Accessories/Eyepieces/ (accessed 07/03/2025).

52 Celestron® Eyepiece Specifications, 2023 https://www.celestron.com/blogs/knowledgebase/how-do-i-use-my-telescope-eyepieces (accessed 07/03/2025).

53 Omegon Plössl eyepieces https://www.omegon.eu/telescope-accessories/eyepieces/15_10/m,Omegon/a,Zubehoer.Allgemein.Serie=Pl%C3%B6ssl (accessed 08/03/2025).

54 Omegon super Plössl eyepiece https://www.omegon.eu/telescope-accessories/eyepieces/15_10/m,Omegon/a,Zubehoer.Allgemein.Serie=Super+Pl%C3%B6ssl (accessed 08/03/2025).

55 Super Plössl, Skywatcher website https://skywatcher.com/product/super-pl-10mm/ (accessed 06/03/2025).

56 TPL eyepiece datasheet, Takahashi https://takahashi-europe.com/catalog/accessories/eyepieces/tpl-eyepieces/tpl-eyepieces (accessed 06/03/2025).

57 Tele Vue® Eyepieces http://televue.com (accessed 07/03/2025).

58 Vixen® Telescope Vixen Premium Eyepiece NPL Series of 31.7 mm eyepiece https://global.vixen.co.jp/en/product/39201_8/ (accessed 07/03/2025).

59 Edmund Optics RKE® Precision Eyepieces, https://www.edmundoptics.co.uk/f/edmund-optics-rke-precision-eyepieces/12484/ (accessed 27/02/2025).

60 *Sold*, Onyx Asset Advisors' website https://thinkonyx.com/sale/optronic-holding-corp/ (accessed 08/03/2025).

61 R. Kingslake and R.B. Johnson, *Lens Design Fundamentals*, 2nd edition, Academic Press, 2010, Chapter 12.

62 H. Erfle, Ocular, United States Patent Office, US1478704, 1925 https://worldwide.espacenet.com/patent/search/family/023954998/publication/US1478704A?q=pn%3Dus1478704 (accessed 26/2/2025).

63 M.J. Kidger, *Fundamental Optical Design*, SPIE, 2000, Chapter 11.

64 R. Kingslake and R.B. Johnson, *Lens Design Fundamentals*, 2nd edition, Academic Press, 2010, Chapter 16.

65 R. Kingslake and R.B. Johnson, *Lens Design Fundamentals*, 2nd edition, Academic Press, 2010, Chapter 16.

66 M.J. Kidger, *Fundamental Optical Design*, SPIE, 2000, Chapter 11.

67 F. Altman, Eyepiece, United States Patent Office, US2423676, 1947 https://worldwide.espacenet.com/patent/search/family/024033298/publication/US2423676A?q=us2423676 (accessed 03/03/2025).

68 J.R. Miles, Binocular eyepiece, Canadian Intellectual Property Office, CA439855, 1947 https://worldwide.espacenet.com/patent/search/family/035555089/publication/CA439855A?q=pn%3DCA439855A (accessed 26/2/2025).

69 H. Erfle, Ocular, United States Patent Office, US1478704, 1925 https://worldwide.espacenet.com/patent/search/family/023954998/publication/US1478704A?q=pn%3Dus1478704 (accessed 26/2/2025).

70 A. König, Telescope eyepiece, United States Patent Office, US2206195, 1940 https://worldwide.espacenet.com/patent/search/family/007990041/publication/US2206195A?q=pn%3DUS2206195A (accessed 28/02/2025).

71 R.B. Tackaberry and R.M. Muller, Telescope eyepiece system, United States Patent Office, US2829560, https://worldwide.espacenet.com/patent/search/family/024466833/publication/US2829560A?q=pn%3DUS2829560A (accessed 04/03/2025).

72 M. Ludewig, Eyepiece for optical instruments, United States Patent Office, US2637245, 1953 https://worldwide.espacenet.com/patent/search/family/007996300/publication/US2637245A?q=pn%3DUS2637245A (accessed 26/02/2025).

73 W.H. Scidmore, Wide angle eyepiece, United States Patent Office, US3390935, 1968 https://worldwide.espacenet.com/patent/search/family/023775034/publication/US3390935A?q=pn%3DUS3390935A (accessed 04/03/2025).

74 A. Nagler, Wide angle eyepiece, United States Patent Office, US4525035, 1985 https://worldwide.espacenet.com/patent/search/family/024271160/publication/US4525035A?q=pn%3DUS4525035 (accessed 04/03/2025).

75 M.H. Freeman and C.C. Hull, *Optics*, 11th edition, Butterworth Heinemann, 2003, Chapter 6.

76 M. Yanari, Wide-field eyepiece lens, United States Patent Office, US5774270, 1998 https://worldwide.espacenet.com/patent/search/family/015974549/publication/US5774270A?q=pn%3DUS5774270A (accessed 12/03/2025).

77 Antares® eyepieces https://antares.space/eyepieces (accessed 07/03/2025).

78 Celestron® Eyepiece Specifications, 2023 https://www.celestron.com/blogs/knowledgebase/how-do-i-use-my-telescope-eyepieces (accessed 07/03/2025).

79 Celestron® X-Cel® LX Series https://www.celestron.com/collections/x-cel-lx-eyepieces (accessed 06/03/2025).

80 J.R. Dire, The Masuyama 1.25″ 53° Premium Eyepieces https://www.hutech.com/file/Masuyama%20Eyepieces.pdf (accessed 08/03/2025).

81 Omegon UWA https://www.omegon.eu/telescope-accessories/eyepieces/15_10/m,Omegon/a,Zubehoer.Allgemein.Serie=UWA (accessed 08/03/2025).

82 LE eyepiece, SkyWatcher website https://skywatcher.com/product/le-2mm/ (accessed 06/03/2025).

83 LET eyepiece, SkyWatcher website https://skywatcher.com (accessed 30/07/2025).

84 SkyWatcher website https://skywatcher.com/product/w-6mm/ (accessed 06/03/2025).

85 Explore Scientific® 52° Series Waterproof Eyepieces https://www.explorescientific.com/collections/52-series-eyepieces (accessed 07/03/2025).

86 Explore Scientific® 62° Series Waterproof Eyepieces https://www.explorescientific.com/collections/62-series-eyepieces (accessed 07/03/2025).

87 Eyepiece LE 24 mm https://takahashi-europe.com/catalog/accessories/eyepieces/le-eyepieces/eyepiece-le-24-mm?lang=en (accessed 02/03/2025).

88 Eudiascopic ED eyepiece 35 mm https://www.baader-planetarium.com/en/eudiascopic-ed-eyepiece-35-mm-1–25inch.html (accessed 02/03/2025).

89 Baader Planetarium Eudiascopic ED 35mm Eyepiece https://www.harrison-telescopes.co.uk/acatalog/baader-planetarium-eudiascopic-ed-35mm-eyepiece.html (accessed 02/03/2025).

90 Vixen® SLV https://global.vixen.co.jp/en/product/tls1040103/ (accessed 07/03/2025).

91 A. Nagler, Plössl Type Eyepiece for Use in Astronomical Instruments, US Patent 4482217, 1984, https://worldwide.espacenet.com/patent/search/family/023857522/publication/US4482217A?q=pn%3Dus4482217 (accessed 23/02/2025).

92 Tele Vue Optics was founded by Al Nagler in 1977. http://televue.com (accessed 04/03/2025).

93 R. Kingslake and R.B. Johnson, *Lens Design Fundamentals*, 2nd edition, Academic Press, 2010, Chapter 14.

94 M.H. Freeman and C.C. Hull, *Optics*, 11th edition, Butterworth Heinemann, 2003, Chapter 6.

95 A. Taylor and H.D. Taylor, Improvements in Telescopes, Microscopes and the Like, United Kingdom Patent Office, GB166217 https://worldwide.espacenet.com/patent/search/family/009989684/publication/GB166217A?q=pn%3DGB166217A (accessed 06/03/2025).

96 H.D. Taylor, A simplified form and improved type of photographic lens, United Kingdom Patent 22607, 1894 https://worldwide.espacenet.com/patent/search/family/032348670/publication/GB189322607A?q=pn%3DGB189322607A (accessed 06/03/2025).

97 A.A. Levin, Eyepiece for Telescopes, United States Patent Office, US2620706, 1952 https://worldwide.espacenet.com/patent/search/family/021967726/publication/US2620706A?q=US2620706 (accessed 03/03/2025).

98 H. Köhler, Ein neues Fernrohrokular mit extrem großem Sehfeld, *Optik*, 17, 500–509, 1960.

99 W.H. Scidmore and R.J. Wolfe, Wide Angle Eyepiece with Large Eye Relief, United States Patent Office, US3464764, 1969 https://worldwide.espacenet.com/patent/search/family/024904003/publication/US3464764A?q=pn%3DUS3464764A (accessed 05/03/2025).

100 A. Nagler, Ultrawide angle flat field eyepiece, United States Patent Office, US4286844, 1981 https://worldwide.espacenet.com/patent/search/family/022230561/publication/US4286844A?q=US4286844 (accessed 03/03/2025).

101 Nagler's 1981 patent[100] explicitly calls the negative doublet a Barlow lens, recognising its astronomical heritage in a way that Scidmore and Wolfe,[99] working from a military design perspective, did not. Kohler[98] and Takahashi[119] refer to the negative doublet as a Smyth lens, reputedly in recognition of Charles Piazzi Smyth, former Astronomer Royal of Scotland, realising the significance of negative lenses as field flatteners. C.P. Smyth was a contemporary of Peter Barlow; Barlow in turn had worked with John Dolland (see Chapter 4, Note 8) on achromatic doublets, of which the Barlow lens is a negative-power example. However, Taylor and Taylor[95] in 1921 referred to their four-element implementation simply as a negative corrector lens, without naming either Smyth or Barlow as forerunners. Moreover, H. Taylor seems to claim the discovery of the significance of field flattening the Petzval curvature and correcting astigmatism in his 1894 patent[96]; he notes prior work by "as long ago as 1858 a certain scientific man named Mr. Sutton", but goes on to conclude that Mr. Sutton's negative lens, along with those employed by Dallmeyer and Ross, were too weak (<0.4x the required strength) to effect

adequate improvement of the curvature, and that most of the improvement Sutton saw was due to a diaphragm, not the negative lens. Taylor does not mention either Barlow or Smyth, but his endeavour at the time was to work on photographic lenses, not astronomical optics.

102 A. Nagler, Ultra wide-angle eyepiece, United States Patent Office, US4747675, 1988 https://worldwide.espacenet.com/patent/search/family/022001903/publication/US4747675A?q=pn%3DUS4747675A (accessed 04/03/2025).

103 D. Dilworth, Extreme wide angle eyepiece with minimal aberrations, United States Patent Office, US4720183, 1988 https://worldwide.espacenet.com/patent/search/family/025265749/publication/US4720183A?q=pn%3DUS4720183A (accessed 12/03/2025).

104 S. Sugawara, Eyepiece lens of wide visual field, United States Patent Office, US5684635, 1987 https://worldwide.espacenet.com/patent/search/family/027321729/publication/US5684635A?q=pn%3DUS5684635A (accessed 12/03/2025).

105 Yang Xihua, Super Wide Angle Eyepiece, Chinese Patent CN208060839, 2018 https://worldwide.espacenet.com/patent/search/family/063984102/publication/CN208060839U?q=pn%3DCN208060839 (accessed 17/03/2025).

106 Ruisi Fu, Ultra wide-angle optical eyepiece system, Chinese Patent CN202305995, 2012 https://worldwide.espacenet.com/patent/search/family/046374863/publication/CN202305995U?q=pn%3DCN202305995 (accessed 17/03/2025).

107 T.L. Clarke, Simple flat-field eyepiece, *Applied Optics*, 22(12), 1807–1811 (1983) https://doi.org/10.1364/AO.22.001807 (accessed 19/04/2024).

108 Hyperion datasheet, Baader Planetarium https://www.baader-planetarium.com/en/downloads/dl/file/id/92/hyperion-eyepieces-technical-data.pdf (accessed 05/03/2025).

109 Hyperion® 68° datasheet, Baader Planetarium https://www.baader-planetarium.com/en/downloads/dl/file/id/155/hyperion-68-eyepieces-brief-description-and-recommended-use.pdf (accessed 05/03/2025).

110 W.H. Scidmore and R.J. Wolfe, Wide Angle Eyepiece with Large Eye Relief, United States Patent Office, US3464764, 1969 https://worldwide.espacenet.com/patent/search/family/024904003/publication/US3464764A?q=pn%3DUS3464764A (accessed 05/03/2025).

111 Telescope accessories Series Ultima Edge, https://www.celestron.com/collections/telescope-accessories (accessed 06/03/2025).

112 Explore Scientific® 68° Series Waterproof Eyepieces https://www.explore-scientific.com/collections/68-series-eyepiece (accessed 07/03/2025).

113 Meade® Series 5000 HD-60 https://www.meadeuk.com/Meade-Series-5000-HD-60-Eyepieces.html (accessed 08/03/2025).

114 Omegon Cronus eyepiece https://www.omegon.eu/telescope-accessories/eyepieces/15_10/m,Omegon/a,Zubehoer.Allgemein.Serie=Cronus (accessed 08/03/2025).

115 Omegon Redline SW https://www.omegon.eu/telescope-accessories/eyepieces/15_10/m,Omegon/a,Zubehoer.Allgemein.Serie=Redline (accessed 08/03/2025).

116 Omegon LE series https://www.omegon.eu/telescope-accessories/ eyepieces/15_10/m,Omegon/a,Zubehoer.Allgemein.Serie=Super+LE (accessed 08/03/2025).

117 Omegon LE Planetary series https://www.omegon.eu/telescope-accessories/ eyepieces/15_10/m,Omegon/a,Zubehoer.Allgemein.Serie=LE+Planetary (accessed 08/03/2025).

118 Pentax XW series https://pentax.eu/products/sport-optics-eyepiece-smc-pentax-xw-10#spec (accessed 09/03/2025).

119 TOE Eyepieces, Takahashi website https://takahashi-europe.com/catalog/ accessories/eyepieces/toe-eyepieces/toe-eyepieces (accessed 06/03/2025).

120 See Note 101.

121 Tele Vue® Eyepieces http://televue.com (accessed 07/03/2025).

122 Vixen® LVW series https://global.vixen.co.jp/en/product/3727_06/ (accessed 07/03/2025).

123 R.B. Rabbetts, *Bennett and Rabbetts' Clinical Visual Optics*, 4th edition, Butterworth Heinemann, 2007, Chapter 8.

124 D.A. Atchison and G. Smith, *Optics of the Human Eye*, 2nd edition, CRC Press, 2000, Chapter 1.

125 Tele Vue Optics was founded by Al Nagler in 1977. http://televue.com (accessed 04/03/2025).

126 Tele Vue® Eyepieces http://televue.com (accessed 07/03/2025).

127 Antares® eyepieces https://antares.space/eyepieces (accessed 07/03/2025).

128 Morpheus® 72° eyepiece series datasheet, Baader Planetarium https://www. baader-planetarium.com/en/downloads/dl/file/id/223/morpheus-76-eye-piece-series-technical-data.pdf (accessed 05/03/2025).

129 Telescope accessories Series Luminos eyepieces, Celestron® website https://www.celestron.com/collections/telescope-accessories (accessed 06/03/2025).

130 Explore Scientific 82°Series Waterproof Eyepieces https://www.exploresci-entific.com/collections/82-series-eyepiece (accessed 07/03/2025).

131 Explore Scientific 92°Series Waterproof Eyepieces https://www.exploresci-entific.com/collections/92-series-eyepieces (accessed 07/03/2025).

132 Explore Scientific 100°Series Waterproof Eyepieces https://www.exploresci-entific.com/collections/100-series-eyepiece (accessed 07/03/2025).

133 Explore Scientific 120°Series Waterproof Eyepieces https://www.exploresci-entific.com/collections/120-series-eyepiece (accessed 07/03/2025).

134 J.R. Dire, The Masuyama 1.25″ 53° Premium Eyepieces https://www.hutech. com/file/Masuyama%20Eyepieces.pdf (accessed 08/03/2025).

135 Omegon OGDO series https://www.omegon.eu/telescope-accessories/ eyepieces/15_10/m,Omegon/a,Zubehoer.Allgemein.Serie=Ogdo (accessed 08/03/2025).

136 Omegon Oberon eyepieces https://www.omegon.eu/telescope-accessories/ eyepieces/15_10/m,Omegon/a,Zubehoer.Allgemein.Serie=Oberon (accessed 08/03/2025).

137 Omegon Panorama II eyepiece https://www.omegon.eu/telescope-accessories/eyepieces/15_10/m,Omegon/a,Zubehoer.Allgemein.Serie=Panorama+II (accessed 08/03/2025).

138 Panorama 82° ultrawide angle eyepiece set, SkyWatcher website https://skywatcher.com/product/panorama-eyepieces/ (accessed 06/03/2025).

139 Nirvana™-ES UWA-82° Eyepiece https://www.opticalvision.co.uk/astronomical_accessories-eyepieces.html (accessed 30/07/2025).

CHAPTER 6

Beyond the Spherical Lens

6.1 GHOSTBUSTING

Fresnel Reflection

Eyepieces vary not only in their lens design and how well they control aberrations, as examined in Chapters 3–5, but also in how well they suppress stray light. Stray light has several origins.

Every refracting surface involves a change in refractive index, and whenever there is a step change in refractive index, light will be partially reflected depending on the square of the difference between the refractive indices of the two media. This is called Fresnel reflection.[1] For light incident at an air-glass (or glass-air) surface at 0° angle of incidence, the fraction of the incident light intensity reflected is $R_{0°} = \dfrac{(n_g - n_a)^2}{(n_g + n_a)^2}$. For a typical air-glass surface with refractive indices $n_g = 1.5$ and $n_a = 1.0$, we expect $R_{0°} = 0.04$, i.e. 4%. At high angles of incidence approaching 90°, the reflected fraction approaches 100%, and under that condition just about any smooth surface, including a smooth black surface, becomes an excellent reflector of grazing-incidence light.

These two limits present three different problems for eyepiece design:

- Reflection losses at normal incidence mean that we stand to lose 8% (2 × 4%) of the light for each lens component it passes through, and for an eyepiece comprising two or more components, those losses can mount up, and the image gets fainter.

 DOI: 10.1201/9781003670506-6

- Light that is reflected doesn't vanish, it just goes elsewhere ... and if it is reflected a second time, it may be redirected towards the observer's eye again and build up an unfocussed or poorly focussed background light level that diminishes contrast in the intended image. The appearance of unwanted defocussed light in an image is called a ghost.

- Light from off-axis sources that strike the inner surfaces of the eyepiece barrel at near-grazing incidence can also be directed towards the observer's eye where it will diminish the contrast in the intended target. The same can occur for light striking the edges of lenses.

Antireflection Coating

The reflection of light from air-glass surfaces can be reduced by depositing a very thin layer of material of intermediate refractive index (between that of air and glass) and of the correct thickness ($\lambda/4$) to set up the reflection of two wavefronts separated by $\lambda/2$, one reflected at the air-coating surface and the second reflected at the coating-glass surface. At normal incidence, the two reflected wavefronts will therefore be half a wavelength apart and will cancel one another out; cancellation will be significant but imperfect at other angles of incidence and at other wavelengths. The phase separation of the two reflected wavefronts depends, of course, on the wavelength of the light, so antireflection coatings typically give a lens surface a coloured cast.

The original antireflection coating developed by Smakula[2] and patented for Carl Zeiss[3] in 1939 was calcium fluoride, though the possibility of building up multilayer coatings was described at the outset. Magnesium fluoride (MgF_2) became the classical single-layer coating, but optical technology has advanced[4] with the provision of a wide range of multilayer coatings, typically consisting of 2–3 layers of different materials, that provide lower reflectance over a wider range of wavelengths, as for example[5] shown in Figure 6.1. These are usually called broadband antireflection (BBAR) coatings. Whereas a single-layer MgF_2 coating provides a single U-shaped reflectance curve, a double-layer coating generates two such troughs separated in wavelength, with a weak peak between them, thus providing a W-shaped reflectance curve. The central weak reflectance peak is usually in the middle of the visible spectral range and may give lenses with W-coatings a greenish cast.

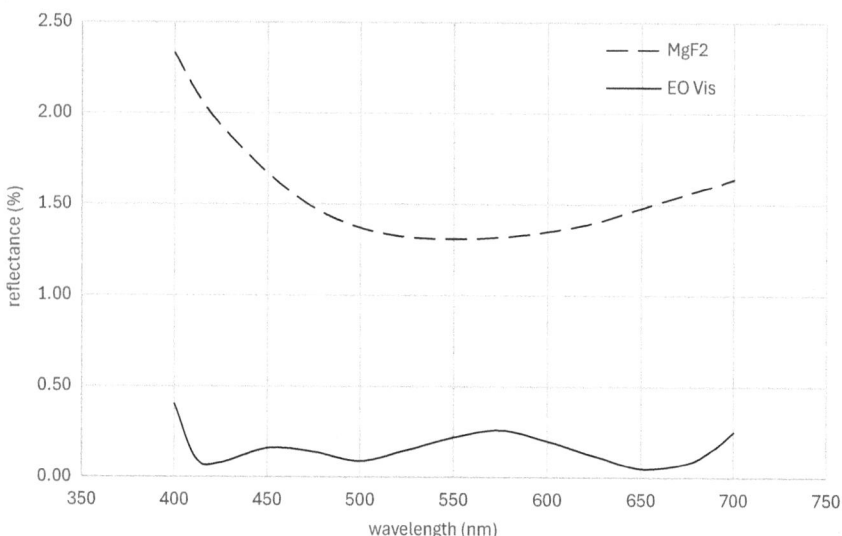

FIGURE 6.1   Reflectance of two antireflection coatings. Upper curve = classic $\lambda/4$ MgF_2; Lower curve = Edmund Optics® Vis 0° multilayer coating.

Besides trying to reduce reflections with antireflection coatings, optical designers will undertake ray tracing of potential ghost-forming reflections to identify which surface combinations are the most troubling and seek to minimise the formation of ghosts as part of the optical design process.

Controlling stray reflections of light from other internal surfaces of the eyepiece barrel requires different strategies, which can include blackening, baffling and roughing of the surface to try to prevent reflected ray paths from reaching the observer's eye, either directly or indirectly.

All of these measures are taken to some degree by eyepiece manufacturers, and some optical designs lend themselves to better stray light control. The rise in popularity of the 19th century Plössl eyepiece in the late 20th century was facilitated in part by the development of antireflection coatings during the 20th century.

The Eye of the Beholder

Given the wide range of optical designs (Chapter 5) and stray light countermeasures that can be adopted by manufacturers, the buyer is faced with a potentially wide range in outcomes, which often cannot be deduced from marketing information. There is much to be said for supplementing technical and marketing information with objectively written reviews of different eyepieces by expert, independent observers.[6] Of course, not all reviews meet those criteria, and sometimes subjective reviews are all that are available.

6.2 ZOOM EYEPIECES

So far in this book, I have described eyepieces having a specified, fixed focal length. Amateur astronomers have predominantly opted for such eyepieces on the basis that they probably only needed around three eyepieces to cover the range of observations they might make with a particular telescope, and there was perceived advantage in having each one optimised to its task. For argument's sake, a small eyepiece collection could feature a short focal-length orthoscopic for high-magnification planetary or double star observations of targets < 1 arcmin across, a mid-focal-length Plössl or similar eyepiece for intermediate magnifications of targets from 1 to 10 arcmin across, and a very wide-field eyepiece for larger star clusters and other extended targets that benefit from accessing the maximum field of view the telescope can deliver. However, zoom eyepieces are also available that allow the observer to vary the focal length – and hence magnification – of the eyepiece to suit the object and seeing conditions, often with a fixed apparent field of view. We saw already in Section 5.7 that a Barlow lens effectively reduces the focal length of an eyepiece by a factor that depends on its position, so repositioning a Barlow lens while in use adjusts the Barlow factor and hence the effective focal length of the eyepiece. Below we examine two other schemes for the construction of zoom lenses generally; neither is especially well suited to eyepieces, but they serve to illustrate the principles by which lens systems of variable focal length can be constructed, and some of the constraints facing such schemes.

Two-Element Thin-Lens Layout

The two-element thin-lens zoom eyepiece about to be described serves principally to provide some insights into strategies for developing zoom eyepiece systems, but it is an oversimplified design. We saw in Section 2.5 the equation for the effective power of a pair of separated thin lenses, viz. $F_E = F_1 + F_2 - dF_1F_2$ (Equation 2.17). It is evident from this equation that once we have chosen a pair of thin lenses, we can vary the effective focal power of the combination, and hence the effective focal length, by varying their separation d. If they are in contact, i.e., if $d = 0$, then the effective power is just their sum $F_1 + F_2$, but if we separate them, the power and focal length change. For a positive-power eyepiece, either one or both lenses must be positive. If both are positive, then the term dF_1F_2 will likewise be positive and the effective power will be reduced (i.e. focal length extended) by increasing d. How much the focal length is extended will depend on the product F_1F_2 and how far apart they are moved, i.e. d. (The alternative possibility, having one lens positive and the other negative, is akin to an adjustable form of the two-component telephoto lens layout already discussed in Section 2.6, or an adjustable Barlow.)

Having decided to pair two positive lenses, the greatest effective power (shortest effective focal length) will be when they are in contact, and the power will reduce (the effective focal length will increase) as they are separated. We will use subscript 1 for the eye lens, and 2 for the field lens (though it doesn't actually serve as a field lens in this application). We do not want the field lens to sit at the focal point of the eye lens as it would then be in focus with the image, as would any defects on it, so we can afford to vary d from 0 up to almost f_1'. When d reaches $f_1'(=1/F_1)$, the separated thin-lens equation tells us that the effective power has reduced to $F_E = F_1$, from $F_1 + F_2$ when in contact. Consequently, the focal power of this pair can be varied over a factor $R \equiv F_{max}/F_{min} = (F_1 + F_2)/F_1$.

We cannot position lens 2 further from the eye lens than the distance f_2', as doing so would place the exit pupil of the telescope+field lens combination inside the eyepiece. So, the maximum separation of the eye lens and lens 2 will necessarily be the lesser of f_1' and f_2'. For a given maximum power $F_1 + F_2$, and corresponding minimum focal length, we can maximise the permissible range of d by making F_1 and F_2 equal. The maximum range factor of this 2-element, thin-lens design is therefore $R = 2$.

If we have in mind that we would like a zoom eyepiece from, say, 12.5 to 25 mm focal length, then that requires $F_{max} = F_1 + F_2 = 80$ D and $F_{min} = F_1 = 40$ D, hence $F_2 = 40$ D also. In practice, we wouldn't let d reach f_1' (or f_2') so we won't quite reach $F_E = 40$ D (25 mm focal length), but thin lenses would come close.

We can now calculate the necessary lens locations relative to the telescope focal plane for these two extreme settings. For the lens pair set in contact ($d = 0$) giving 80 D and thus $f_E' = 12.5$ mm, and recalling we always observe in infinity adjustment, the pair of thin lenses in contact must be located 12.5 mm from the telescope focal plane. When they are moved to their maximum separation, with lens 2 essentially at the focal point of the eye lens, we realise that lens 2 is now coincident with both the focal plane of the eye lens and necessarily also with the telescope focal plane; lens 2 is now in the role of an ideal field lens where it no longer contributes to the focal power of the eyepiece (which is why $F_E = F_1$). The eye lens in this case is now a distance f_1' from the telescope focal plane, i.e. 25 mm.

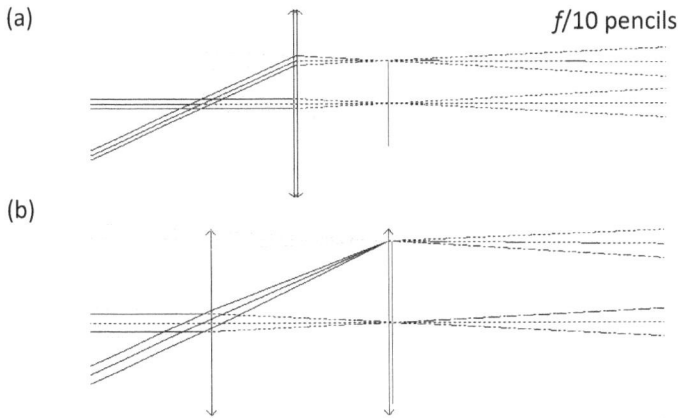

FIGURE 6.2 Thin-lens layouts for a hypothetical two-element, 12.5–25 mm, zoom eyepiece, for 0° and 24° field angles. The thin lenses are represented by long vertical lines with arrow heads top and bottom, while the telescope focal plane is represented by a short vertical line. $f_1' = f_2' = 25$ mm. All lens diameters (heights in the diagram) have been set to 25 mm as a consistent scale. (a) $f' = 12.5$ mm. (b) as for (a) but adjusted to $f' = 25$ mm.

The two-element zoom eyepiece described above poses a moderate challenge for an observer, evident by comparing Figure 6.2a and b. Although the exit pupil has remained stationary 25 mm behind the telescope focal plane as the zoom factor is varied, the eye lens moves in and out by a distance equivalent to the focal range, in this case 12.5 mm, and this could be disconcerting for an observer in the dark. There may be a preference amongst observers to have the eye lens stationary relative to the telescope focal plane, with the change of focal length effected by moving internal lens elements other than the eye lens. We explore this possibility below.

Three-Element Photographic Zoom Lens

One photographic zoom lens with properties like those just described uses a positive-negative-positive three-lens layout.[7] The positive lens facing the distant object (infinite conjugate) is a fixed distance from the focal plane of the lens system (finite conjugate), while the other two lenses are moved closer together and further apart to adjust the internal magnification of the system and hence its effective focal length. The conjugates in this case match those for an eyepiece: the distant conjugate is the distant image formed by infinity adjustment of a telescope+eyepiece combination, and the finite conjugate represents the real image in the focal plane of the telescope. However, this similarity is deceptive. It was noted at the outset of this book (Section 1.1) that one of the significant distinguishing features of eyepieces compared to other optical systems is the external location of the aperture stop, in contrast to a telescope objective or camera lens where the aperture stop is generally co-located with the optics. The externality of the aperture stop, and thus also the entrance and exit pupils of an eyepiece, introduces a significant constraint on design; it is also the reason that the eyepiece ray traces must include an aperture stop located at an assumed telescope exit pupil, to ensure that the constraints it imposes of the ray paths in the eyepiece are obvious. With a positive-negative-positive three-lens photographic zoom lens, the designer can locate the aperture stop wherever they want it inside the lens assembly; some choices are obviously better than others. However, the entrance pupil of an eyepiece is either the telescope aperture (for refractors and Newtonian reflectors) or its image (for a Cassegrain-style telescope possessing a convex secondary mirror), in both cases located well outside the eyepiece. One consequence is that the exit pupil of the telescope+eyepiece combination cannot be allowed to fall inside the eyepiece assembly, where it would be inaccessible to the observer's eye. This means the three-element photographic zoom lens layout cannot trivially be rescaled to smaller dimensions for use as an eyepiece; the external location of the aperture stop creates a significant constraint.

While neither the two-element thin-lens layout nor the three-element photographic zoom lens layouts work especially well for eyepieces, they nevertheless illustrate the possibility of varying the focal length of a lens system by varying the relative positions of some of its lenses.

Commercial Eyepieces

The traditional objection of amateur astronomers to using zoom eyepieces has been that the convenience of such an eyepiece does not necessarily make up for the design trade-offs required to achieve that capability, compared to a small selection of individually optimised, fixed-focus eyepieces potentially of a similar price. However, a number of zoom eyepiece designs have been commercialised, including from some noted designers. They fall into a few relatively restricted ranges.

A number of zoom eyepieces concentrate on a narrow range of short focal lengths typically associated with high-magnification observations, intended for observers seeking to tune their magnification of detailed features to the seeing conditions. These include:

- 3–6 mm: Tele Vue® Nagler™ Planetary Zoom[8] (1¼ inch barrel), with a 50° apparent field of view and 10 mm eye relief.

- 3–8 mm: SVBONY Planetary Zoom Eyepiece[9] (1¼ inch barrel), using a 6-element, 4-group design with a 56° apparent field of view.

- 7.7–15.4 mm: APM Super Zoom[10] (1¼ and 2 inch barrel fittings), providing a 66–67° apparent field of view.

A slightly longer focal-length range of around 8–24 mm is covered by a wider range of manufacturers including:

- 7–21 mm: Neewer LS-T22[11] (1¼ inch barrel), 5-element, 3-group eyepiece providing 15 mm eye relief, 40°–57° apparent field of view, and weighing 157 g.

- 7.2–21.5 mm: Lunt LS7–21ZE[12] (1¼ inch barrel) 7-element, 4-group eyepiece providing 40°–53° apparent field of view, 15 mm eye relief and weighing 180 g.

- 8–20 mm: SVBONY SV230 Super Zoom Aspheric[13] (1¼ and 2 inch barrel fittings), 9-element, 6-group eyepiece giving a 57°–72° apparent field of view and 17–19 mm eye relief, weighing 500 g.

- 8–24 mm: Baader Mark IV Hyperion[14] (1¼ and 2 inch barrel fittings), a 7-element, 4-group eyepiece providing a 68° apparent field of view, weighing in at 290 g.

- 8–24 mm: Bresser LER,[15] 9-element, 6-group eyepiece, 40°–62° apparent field of view, and 20 mm eye relief, at 315 g.

- 8–24 mm: Celestron® Zoom Eyepiece[16] (1¼ inch barrel), 40°–60° apparent field of view, and 15–18 mm eye relief.

6.3 ASPHERICS

Throughout the preceding chapters, we have adopted lens surfaces that are spherical, for two reasons: (1) they have a geometry that is easy to describe mathematically (Section 2.3–2.4) which facilitates a relatively straightforward process for calculating ray paths, and (2) they are easier to manufacture and test than other shapes, which makes them more affordable to commercialise in glass, though optical plastics can allow aspherics to be made more economically than in the past. However, we have seen that spherical surfaces have drawbacks, principally that a spherical wavefront entering an optical system will almost inevitably become non-spherical as a result of passing through spherical optical surfaces, giving rise to aberrations and poorer image quality. The prospect of avoiding or offsetting wavefront aberrations by introducing an aspheric optical surface into a system is one that cannot be ignored.

The theory of aspherics in optics is not new; Descartes described the properties of a plano-hyperbolic lens, which we shall meet below, in the 1600s, but manufacturing technology was too primitive then to allow the design to be realised in glass.[17] Many astronomers will already be familiar with the adoption of aspheric surfaces in the primary optics of most reflecting telescopes, such as the parabolic (paraboloidal) primary mirror and hyperboloidal secondary mirror of a Cassegrain telescope, and the aspheric corrector plate of a Schmidt-Cassegrain. So, what of aspheric surfaces in eyepieces? We begin by looking at aspheric shapes, some realisations in lenses and their adoption in eyepieces.

Mathematical Description of Aspherics

Taken literally, "aspherical" means not spherical, but in optics, the term is usually restricted to a category of surfaces that are rotationally symmetric; the axis of symmetry ultimately will be the optical axis of the system into which the surface is integrated. Secondly, while a Schmidt-Cassegrain corrector plate has a "wavy" shape which is the sum of quadratic and quartic functions,[18] many aspherics of interest will be 3D conic surfaces. A 2D conic is a mathematically well-described curve corresponding to a cross-sectional slice through a cone, i.e. an ellipse (one form of which is a circle), parabola or hyperbola. Rotation of these curves about their axis of symmetry generates a 3D conic surface; spherical surfaces represent one specialist case of the wider class of possible conic surfaces.

We saw in Figure 2.4 that the profile of a surface can be described by its sag, given mathematically as the relationship between the distance of a surface point from the optical axis, i.e., its centration y, and its distance ζ along the optical axis from the vertex, with $\zeta = r - \sqrt{r^2 - y^2}$ (Equation 3.2). With a little algebra,[19] we can also write this as

$$\zeta = \frac{y^2}{r\left(1 + \sqrt{1 - \dfrac{y^2}{r^2}}\right)} \tag{6.1}$$

It can be convenient to divide Equation 3.2 by r and write $\frac{\zeta}{r} = 1 - \sqrt{1 - \left(\frac{y}{r}\right)^2} = 1 - \left(1 - \left(\frac{y}{r}\right)^2\right)^{1/2}$. This form is useful because on a spherical surface, the centration y is always less than the radius of curvature r, so we can expand the right-hand side using the binomial expansion to $\frac{\zeta}{r} = 1 - \left(1 - \frac{1}{2}\left(\frac{y}{r}\right)^2 - \frac{1}{8}\left(\frac{y}{r}\right)^4 - \frac{1}{16}\left(\frac{y}{r}\right)^6 - \ldots\right)$, thus giving

$$\zeta = \frac{1}{2}\frac{y^2}{r} + \frac{1}{8}\frac{y^4}{r^3} + \frac{1}{16}\frac{y^6}{r^5} + \ldots \tag{6.2}$$

Note that this equation is general to a spherical form, whether a spherical glass surface or a spherical wavefront.[20] Furthermore, if the centration y is very small compared to the radius of curvature r, the term in y^4 can be ignored along with higher-order terms, in which case the sag of the spherical surface appears to have a quadratic dependence on centration y, given by $\zeta \approx \frac{y^2}{2r}$. This relationship indicates that a small portion of a spherical surface is indistinguishable from a shallow paraboloid. This reality is sometimes exploited as an economy in inexpensive reflecting telescopes where the primary mirror may be implemented in a spherical rather than paraboloidal form. This economy isn't as shocking as it might sound; a very small telescope operating at $f/10$ or slower may have an Airy disk larger than the circle of least confusion arising from spherical aberration.[21,22]

In fact, we have already met the equation $\zeta = \frac{1}{2}\frac{y^2}{r} + \frac{1}{8}\frac{y^4}{r^3} + \ldots$ in Section 3 where the sag of a spherical wavefront was analysed and approximated under third-order (Seidel) theory, as a first step towards calculating the aberrations due to a non-spherical wavefront. For the wavefront error, we introduced an additional term in y^4 treated as a change in the sag $\Delta\zeta$, and used the symbol ρ rather than y for the centration, writing $\Delta\zeta \equiv_s A\rho^4$. The choice of centration symbol ρ recognised that the ray need not lie in the meridional plane like y, e.g. it might be offset by a distance x in front of or behind the meridional plane, and hence the true centration was given by ρ where $\rho^2 = x^2 + y^2$.

If we consider an optical surface that is instead ellipsoidal or hyperboloidal,[23] we have different relationships between centration ρ and sag ζ, given by $\frac{(\zeta - b)^2}{b^2} + \frac{\rho^2}{a^2} = 1$ for the ellipsoid with semi-major axis a and semi-minor axis b, and $\dfrac{(\zeta-a)^2}{a^2} - \dfrac{\rho^2}{b^2} = 1$ for the hyperboloid. For the ellipsoid, a bit of algebra provides an expression for the sag recognisable as similar to that for the sphere, but with a modification by a new variable ε:

$$\zeta = \frac{\rho^2}{r\left(1 + \sqrt{1 - \varepsilon\dfrac{\rho^2}{r^2}}\right)} \tag{6.3}$$

where $\varepsilon \equiv 1 - e^2$ and e is the eccentricity of the ellipse ($\varepsilon > 0$). The trivial case where the eccentricity $e = 0$ yields $\varepsilon = 1$, and ζ in Equation 6.3 is then seen to revert to the earlier form for a sphere (Equation 6.1). For the hyperboloid, a similar form arises but $\varepsilon < 0$. The intermediate case, with $\varepsilon = 0$, clearly gives rise to $\zeta = \rho^2/(2r)$, i.e. a paraboloid.

This general expression for the sag of a conical surface is usually rewritten slightly in optics, using a different constant k in place of ε, where $k \equiv \varepsilon - 1$ and is called the conic constant, hence $\varepsilon = k + 1$. Optics textbooks and ray tracing programmes usually describe a conical surface using the expression

$$\zeta = \frac{\rho^2}{r\left(1 + \sqrt{1 - (1+k)\dfrac{\rho^2}{r^s}}\right)} \tag{6.4}$$

A binomial expansion of this expression inevitably differs slightly from the spherical one above. For a general conical surface,

$$\zeta = \frac{1}{2}\frac{y^2}{r} + \frac{1}{8}(1+k)\frac{y^4}{r^3} + \frac{1}{16}(1+k)^2\frac{y^6}{r^5} + \ldots \tag{6.5}$$

Comparing the binomial expansions for ζ in the sphere (Equation 6.2) and the general conical surface (Equation 6.5), we see that the first parabolic term is unchanged, but the subsequent terms in the expansion now show terms in $(1 + k)$, so to describe an aspheric surface, we need to provide the radius of curvature r which describes the curvature of the equivalent sphere at low centrations, and also provide the conic constant k which indicates how the higher-order terms need to be modified from the spherical case as the centration increases. It is easy to see that in the hyperboloidal case where $\varepsilon < 0$, the term in $(1 + k)$ becomes negative, and a surface with positive radius of curvature r is pulled back towards the vertex at higher centration, which tends to flatten the flanks of a hyperboloid compared to a spherical surface having the same value of r.

This new algebraic expression for conical surfaces (Equation 6.5) provides a mechanism for calculating the coordinates (ζ,ρ) of points on a conical optical surface. Additional higher-order terms, at higher even powers ρ^4, ρ^6, etc., may also be included if a wider range of non-conic aspheric surfaces is desired, with the result that Equation 6.4 may be superseded by the following extension:

$$\zeta = \frac{\rho^2}{r\left(1+\sqrt{1-(1+k)\dfrac{\rho^2}{r^2}}\right)} + a_4\rho^4 + a_6\rho^6 + a_8\rho^8 + a_{10}\rho^{10} +\ldots \quad (6.6)$$

(Obviously, it doesn't matter what symbols and subscripts are used for the coefficients a_4, etc., and others are in use.)

Whereas we derived a set of algebraic equations for exact tracing of rays through spherical surfaces in Section 2.3, the same cannot be done for finding the points at which rays will intersect aspheric surfaces, so numerical approaches must be used.[24] Computer-based ray tracing programmes excel at this sort of work, whereas manual calculation of rays through aspherics would be much more time-consuming. Nevertheless, once the intersection point of a ray and the aspheric surface is determined, Snell's Law can be applied to calculate the new ray trajectory and so on through the system.

Optical Justification for Aspherics

It is all very well having a mathematical description for an aspheric, but why might one want that in the first place, and what additional problems might aspheric surfaces create? We saw in the analysis of spherical surfaces (Chapters 2 and 3) that parallel rays striking a positive lens with spherical surfaces are *not* all brought to a focus; rather marginal rays cross the optical axis at a shorter distance than paraxial rays, which we describe as positive spherical aberration. We can imagine a different situation in which parallel rays refracted by a positive surface *are* brought to a common focus and ask what surface shape that outcome would require. Obviously, if the marginal rays are to cross the optical axis at the same location as the

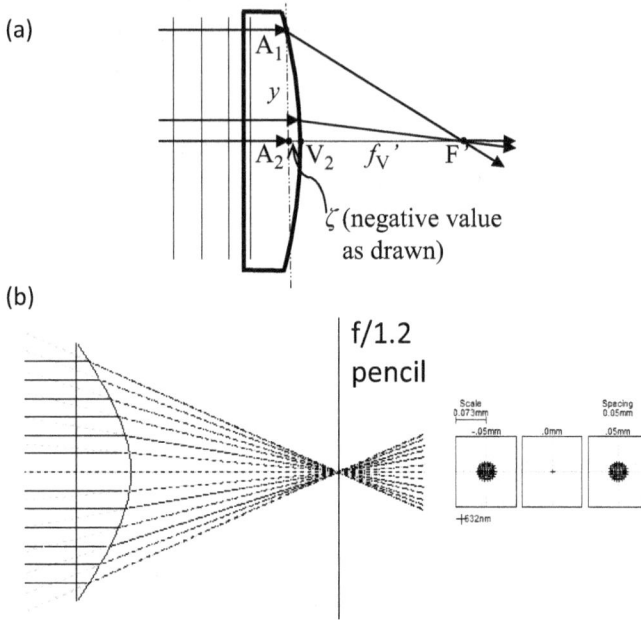

FIGURE 6.3   Aspheric lens which brings all rays of an incident parallel pencil to a common focus at F'. (a) schematic. (b) ray trace and spot diagrams showing exquisite image formation for $f/1.2$ pencils of an on-axis image at the design wavelength of an aspheric lens (plano + hyperboloid; $r = 12.99$ mm, $k = -2.31$, N-BK7, $t = 6.61$ mm).

paraxial rays, the refraction in the flanks of the surface must be weakened compared to the spherical case; we see how below.

In Figure 6.3a, three parallel rays are shown passing through a plano surface of a lens and striking the second surface. One ray lies along the optical axis, a paraxial ray is shown close by, and the third is close to the top edge; we shall call this one the marginal ray since that terminology is familiar. The incoming wavefronts associated with the rays, shown schematically by four vertical lines, slow down when they enter the lens of refractive index n, but as the first surface is plano and the angle of incidence is zero, they remain straight until they reach the plane A_1A_2, at which stage the marginal ray exits the glass and takes the path from A_1 to F' through air. In the same time that the marginal ray reaches F', the ray on the optical axis travels a distance A_2V_2 through glass and a distance V_2F' through air.

The travel time from A_1 to F' for the marginal ray is given by $t_M = \overline{A_1 F'}/v_{air}$ where the distance $\overline{A_1 F'}$ is given by Pythagoras' theorem as $\overline{A_1 F'}^2 = y^2 + (-\zeta + f'_V)^2$. Note that by the Cartesian convention ζ is negative in Figure 6.3 as it is measured to the left of the vertex V_2, so rays travelling a distance $\overline{A_2 V_2}$ from left to right cover a positive distance given by $-\zeta$. Meanwhile, the axial ray takes a time $t_A = \overline{A_2 V_2}/v_{glass} + f'_V/v_{air}$ to reach F', where again the positive distance $\overline{A_2 V_2} = -\zeta$. Both rays must take the same time to reach F' if they are to arrive in phase, i.e. to intersect and produce an image at F', so we can equate the two times $t_M = t_A$ and write $[y^2 + (-\zeta + f'_V)^2]^{1/2}/v_{air} = -\zeta/v_{glass} + f'_V/v_{air}$. Multiplying through by v_{air} simplifies the relation to $[y^2 + (-\zeta + f'_V)^2]^{1/2} = -n\zeta + f'_V$ where n is the refractive index of the glass. The equation is not exactly pretty, but once we specify the desired second vertex focal length of the lens (f'_V) and the refractive index of the glass, we end up with an expression between just two variables, the sag ζ and the centration y, which therefore gives us an equation uniquely specifying the shape of the lens surface.

We can go further. Squaring both sides of the last equation and rearranging it gives a quadratic equation $(n^2 - 1)\zeta^2 - 2(n-1)f'_V\zeta - y^2 = 0$. This is quadratic in ζ and therefore has two solutions for the sag as a function of centration y: $\zeta = \dfrac{1}{n+1}\left(f'_V \pm \sqrt{f'^2_V + \dfrac{n+1}{n-1}y^2}\right)$. We can note the following behaviours of this solution:

- On the optical axis, where $y = 0$, the two solutions for the sag simplify to $\zeta = 0$ and $\zeta = 2f'_V/(n+1)$. The first solution corresponds to the vertex V_2 in Figure 6.3a where the sag is zero. The second solution indicates the vertex of a second curve, which is not an optical surface, displaced a distance $2f'_V/(n+1) \approx 0.8 f'_V$ to the right of the vertex V_2.

- When the centration y is non-zero but considerably smaller than f'_V, the sag can be expanded using a binomial expansion to show that the surface sag is quadratic in the centration, i.e., proportional to y^2, as is also typical of a spherical surface at low centration. In other words, at low centration, there is little difference between spherical shapes and this aspherical shape.

- When the centration y becomes large enough to dominate the term in the square root, which for $n \approx 1.5$ implies $y > f'_V$, the quadratic solution approximates to $\zeta \approx \frac{2}{5} f_V' \pm \frac{2}{\sqrt{5}} y$. This is the formula of a straight-line relationship (in fact two straight lines) between ζ and y, meaning the profile of the lens in the flanks must be linear (uniformly sloping, and no longer curved).

These characteristics match those of a hyperboloid, and indeed it can be shown with further algebra that the equation linking the sag ζ and centration y is indeed that of a hyperbola.[25]

Note, however, that this perfectly focussing conic surface was derived for some very rigid conditions. It is for (1) incoming parallel light rays from a very distant object; (2) for an on-axis object; (3) for a specified refractive index which applies only to a single wavelength. Consequently, we might worry that such a lens would not function well for nearby object locations, could have off-axis aberrations and will suffer chromatic aberration.

We demonstrate off-axis and chromatic aberrations by ray tracing a lens having this form, based on the Edmund Optics® 532 nm hyperbolic aspheric lens[26] but with the lens diameter maximised to show the ray trace for fast ray pencils. At its design wavelength and for an on-axis object at infinity, the lens produces an exquisitely sharp image even for an $f/1.2$ pencil (Figure 6.3b). However, if we try to employ the same lens as a 25 mm focal length eyepiece capable of providing images of off-axis objects at a range of visible wavelengths, it is clear that we have asked too much of it (Figure 6.4). Even at field angles as low as 6°, the off-axis aberrations are large compared to those we have seen previously.

This state of affairs is a reminder that aspherics have a role in ensuring appropriate shaping of wavefronts, but they are not, on their own, a solution to the many compromises that arise in the design of eyepieces. The typical role of aspherics in eyepieces is therefore to contribute one surface in the design, with the role of improving the aberrations that the rest of the train, comprising spherical surfaces, cannot correct.

Eyepieces Incorporating Aspherics

In Section 5.6, we briefly met eyepiece designs by Sugawara[27] (1987), four of which incorporate an aspheric surface. Example 9 of that patent has six elements using five glass types, which have been matched as closely as possible in the simulation shown in Figure 6.5. The rear surface of the field lens is aspheric, with a strong hyperboloidal conic constant ($k = -6.976$), and three additional terms in ρ^4, ρ^6 and ρ^8 (cf. Equation 6.6); the hyperbolic flattening of the flanks of the lens is evident in Figure 6.5a.

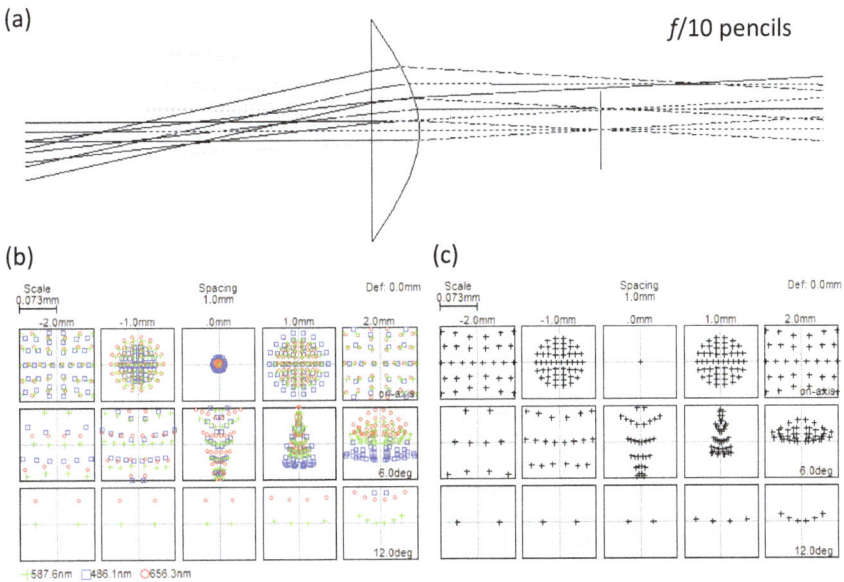

FIGURE 6.4   25 mm focal-length aspheric lens (plano + hyperbola; $r = 12.99$ mm, $k = -2.31$, N-BK7, $t = 6.61$ mm). (a) Ray trace for field angles 0°, 6° and 12°. (b) Spot diagrams for three wavelengths at apparent field angles of 0°, 6° and 12°. (c) Spot diagrams for f/10 pencils at apparent field angles of 0°, 6°, and 12°.

It is apparent quite quickly from Figure 6.5 that the design represents a considerable achievement for a 25 mm focal-length eyepiece, with excellent imaging out to the 24° field angle (the stated apparent field is 59°), minimal pupil spherical aberration and a 19 mm eye relief, all in a relatively compact eyepiece whose axial length (first to last surface) is just 35 mm.

Several eyepiece designs incorporating aspheric surfaces were patented around the turn of the century,[28] with Hall[29] noting that "nearly all the best solutions based on spherical glass optics have already been utilised".

Moriyasu Kanai's designs[30] are of particular note because they demonstrate a level of improvement possible to a fairly simple base design – a Kellner eyepiece. The Kellner's simple positive field lens is replaced by a lens having one (or more) aspheric surface. This obviously counts as a relatively simple eyepiece compared to most of the late 20th century designs where increasing numbers of elements, components and groups were pursued to achieve improved wide-field imaging, at higher cost and higher weight. Kanai also recognised the manufacturing simplicity of using a plastic aspheric lens. So, what did Kanai achieve with

(a)

f/10 pencils

(b)

(c)

(d)

f/6 pencils

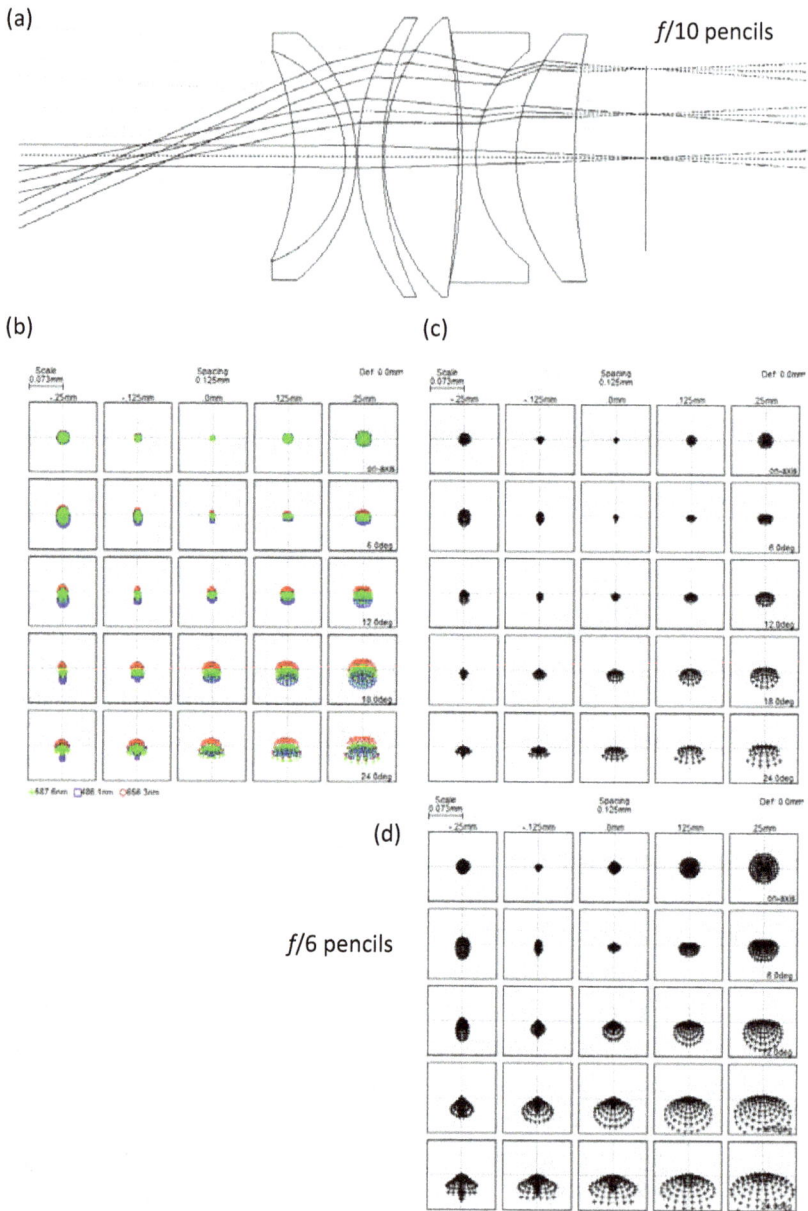

FIGURE 6.5 Eyepiece with an aspherical surface on the rear surface of the field lens [based on Sugawara (1987) Example #9, 25 mm focal length]. (a) Ray trace for field angles 0°, 12° and 24°. (b) Spot diagrams for three wavelengths at apparent field angles of 0°, 6°, 12°, 18° and 24°. (c) Spot diagrams for f/10 pencils at apparent field angles of 0°, 6°, 12°, 18° and 24°. (d) Spot diagrams for f/6 pencils at apparent field angles of 0°, 6°, 12°, 18° and 24°.

an aspherised Kellner? Embodiment 7 of the design is simulated in Figure 6.6, in which the second surface of the field lens is aspherised, not as a hyperboloid but via two higher-order terms (in ρ^4 and ρ^6 with $k = 0$; see Equation 6.6). The spot diagrams are clearly improved relative to spherical Kellners (Section 5.1) sharing the 3-element, 2-group layout, but as expected, it does not achieve as good image quality as the 6-element Sugawara design.

Commercial eyepieces containing aspheric elements are currently not as common as the more traditional eyepieces containing spherical surfaces.

Baader Planetarium markets a 6-element, 4-group "Hyperion Aspheric"[31] eyepiece in 31 and 36 mm focal lengths, providing 72° apparent field of view in a 2 inch barrel, which can be reduced down to 50°–55° with 1¼ inch fittings. The eye relief is 17–19 mm, and the weight is 356–382 g depending on the configuration.

For a time, Celestron® marketed three eyepieces (4, 10 and 23 mm focal length) with one or more aspheric surfaces, and a 62° apparent field of view; they do not currently appear on the Celestron® website, but are available from other dealers bearing the Celestron® name. Similar eyepieces are available from SVBONY.[32] No details of the optical design are given online, so subjective reviews provide the only available information. SVBONY also markets an 8–20 mm Super Zoom Aspheric Eyepiece[33] (see Section 6.2).

6.4 GRADIENT INDEX LENSES

We saw in Section 2.1 that Snell's Law of refraction has its origin in the change in the wavefront slope and hence direction that occurs when a wave passes from one medium to another where the wave speed is different, if it arrives inclined to the interface. We commonly quantify the wave speeds via refractive indices. The change in speed (or refractive index) does not, however, need to be step-like as in an air-glass surface; a gradual change of speed or refractive index in a material likewise changes the slope of the wavefront passing through it, and hence alters the direction of propagation of the wave. It is therefore possible to refract light continuously over a path, rather than abruptly at a boundary, and replicate the refractive role of a lens. A material in which a gradual change in refractive index has been created is called a gradient index (GRIN) material, and a lens created from such a material is called a gradient-index lens, sometimes shortened to GRIN lens.

To replicate the power of a positive lens, a gradient-index material will be structured with a high refractive index along its central axis, which

(a)

$f/10$ pencils

(b)

(c)

(d)

$f/6$ pencils

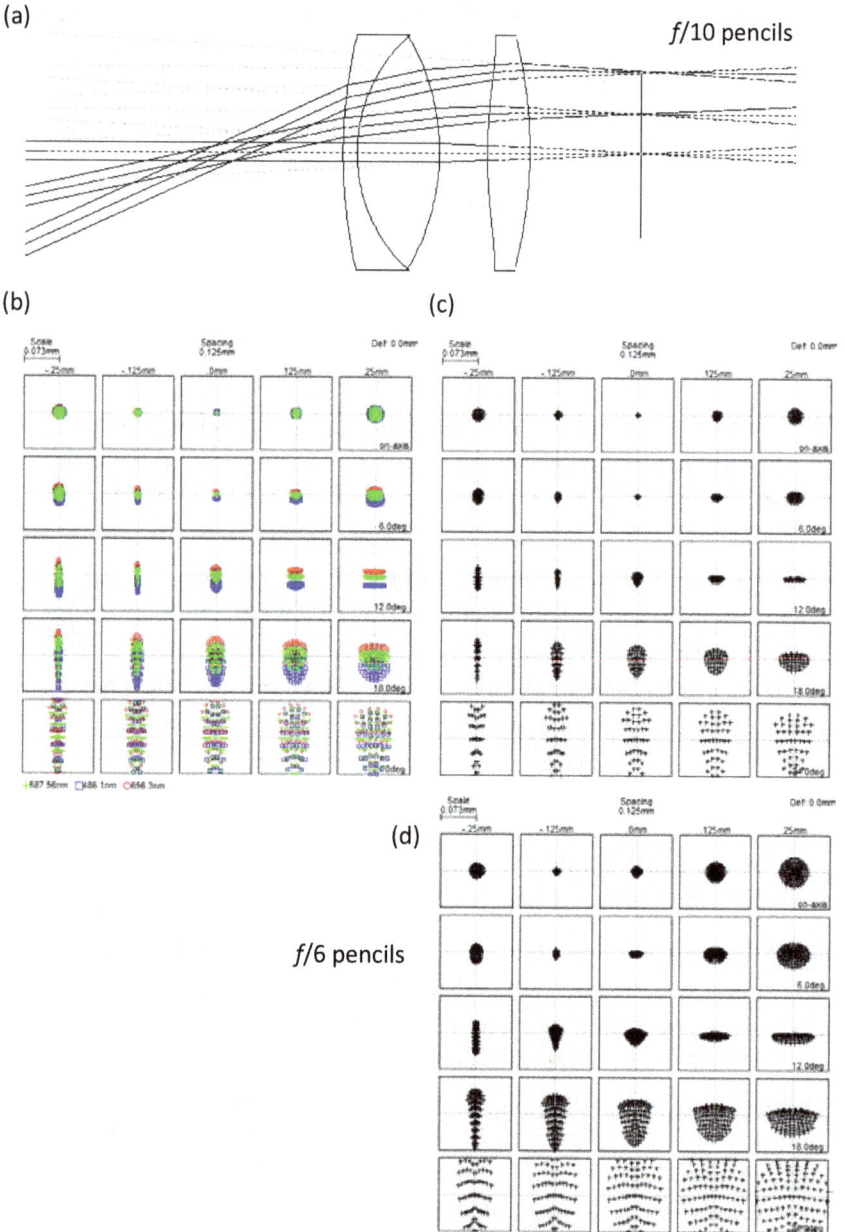

FIGURE 6.6 Eyepiece with an aspherical surface on the rear surface of the field lens [based on Kanai (1987) Embodiment 7, 25 mm focal length]. (a) Ray trace for field angles 0°, 12° and 24°. (b) Spot diagrams for three wavelengths at apparent field angles of 0°, 6°, 12°, 18° and 24°. (c) Spot diagrams for f/10 pencils at apparent field angles of 0°, 6°, 12°, 18° and 24°. (d) Spot diagrams for f/6 pencils at apparent field angles of 0°, 6°, 12°, 18° and 24°.

corresponds to the optical axis, and a decreasing refractive index outwards from the centre. The portion of the wavefront travelling near the outer part of the lens will therefore travel faster than the inner portion and overtake it, hence curving the ray path towards the optical axis. (Recall that rays are the normals to the wavefronts and show their direction of travel.) The variation of refractive index with distance from the optical axis can be modelled by a smoothly varying mathematical function, usually with a quadratic (and possibly higher-order) decrease from the peak value on the axis.[34,35]

Current commercially available gradient-index lenses tend to be for special millimetre-scale applications only,[36,37] such as managing light that will be transmitted through optical fibres, but designs nevertheless exist for photographic lenses and eyepieces utilising macroscopic GRIN lens, and the theory of aberration control in GRIN lenses has likewise been developed.[38,39] It remains to be seen whether these or similar designs will actually be put to use at some future time in eyepiece optics.

6.5 FILTERS, DIAGONALS AND BINOCULAR VIEWERS

Our final topic is to consider three accessories that may be used as optical aids in conjunction with eyepieces. None is essential, but they may improve the observing experience in some situations.

Filters

Filters are thin, flat glass or gelatine disks that can be inserted into the optical path, typically immediately upstream of the telescope focal plane. Their transmission depends on wavelength. They are intended to transmit some chosen wavelengths and suppress others. Physically, they attach to the end of the eyepiece barrel, which typically has a fine internal thread that matches an external thread on the rim of the filter holder.

Filters for visual use fall into three broad categories: neutral density filters which are grey; broadband coloured filters; and specialist background-suppression filters.

Neutral density filters are used simply to reduce the brightness of the object that is being viewed, so the eye is not overwhelmed by an exceptionally bright target, which in practice means the Moon or perhaps the brightest planets. This list explicitly does not include the Sun, for which eyepiece filters should never be used; focussed sunlight can readily cause an eyepiece filter to heat up and crack without warning, so the only safe filter for solar observing is a properly made, highly reflective filter that fits securely and completely over the objective, admitting only a tiny fraction of the light to the telescope.

Neutral density filters are usually graded according to the amount of light they absorb, with larger ND numbers corresponding to more absorption. The fraction of light transmitted is given by $T = 1/10^{ND}$, so an ND0.3 filter transmits $T = 1/10^{0.3} = 0.50$, i.e. 50%, while an ND0.9 transmits $T = 0.13$, i.e. 13%. If you have too much light and want to throw some away, an alternative strategy could be to use a coloured bandpass filter (see below) as this will additionally hide residual chromatic aberration in the telescope or eyepiece (or eye[40,41] for that matter), or from atmosphere dispersion that may be present, thus presenting the eye with a better-focussed mid-wavelength image.

Broadband coloured filters are normally used to increase the contrast of different coloured features of planetary disks, especially Jupiter and Mars and to a lesser degree Saturn and Venus. While it can be possible to *resolve* details on the disks of these planets, the *contrast* of the features to their surroundings can be quite subtle. Jupiter's Great Red Spot provides a good example illustrating the potential of coloured filters. This oval feature of Jupiter's southern equatorial belt region is large, but it is of a similar brightness to surrounding cloud material, and its colour – orangey red – is not so distinct that it stands out easily to the eyes of all observers. However, because it is at least a bit more orangey than the surroundings, a green- or blue-coloured (i.e. green/blue-transmitting) filter will block more of the light from the Great Red Spot than from surrounding regions, making the Great Red Spot appear darker and hence more obvious. Some observers like to use a blue filter to observe Jupiter for this purpose. However, whether you actually *want* Jupiter to look a bit blueish is up to you; you might prefer the natural colours including the subtlety of the different hues, and instead prefer to train your eye+brain combination to become more aware of subtle contrasts; in practice, this means being patient, observing continuously for five minutes or more, and allowing your eye+brain to become accustomed to what is in front of it.

For Mars, a suitable strategy could be to darken the dark (rock) features to make them more obvious against the brighter, red-orange (sandy) zones. For this purpose, a yellow, orange or red filter that will transmit most of the red, but preferentially suppress shorter wavelengths associated with the darker Martian material, may help. To help see the polar caps, however, you might want to suppress the dominant red surface zones by using a green or blue filter; as the bright white light reflected from the polar caps includes a very wide range of wavelengths, a reasonable portion of that light will still be transmitted by a green/blue filter, so they will appear relatively brighter … albeit blue, which is not the natural colour of Mars or its polar caps.

More specialist filters may be used to improve the contrast of deep-sky objects against the sky, which is never completely dark. Didymium is an historical mixture of the two metals praseodymium and neodymium that has an absorption band that coincides extremely well with the bright yellow/orange emission line due to sodium. which is a bright source of emission in hot-worked "soda glass".[42] Didymium can be incorporated into a spectacle glass[43] to form a pale purple-coloured filter which absorbs the sodium emission lines; for around a century, glass-workers have worn didymium eye-glasses as a way of suppressing the bright yellow/orange flare that arises from the sodium in their workplace,[44] while maintaining reasonably normal vision at most other wavelengths.

The absorption spectrum of didymium is principally due to the element neodymium. As sodium has also been present in much street lighting during the late 20th and early 21st century, neodymium filters have been produced for use in astronomy as a means of suppressing light pollution from this particular source.[45] Fortuitously, as the orange light suppressed by neodymium is also a reasonable match to the colour of Jupiter's Great Red Spot, neodymium filters can also double as contrast-enhancers for that planetary feature.[46] With an ongoing shift from sodium to LED street lighting in many countries at present, the relative contribution of sodium to overall light pollution will gradually diminish and be replaced with a broader spectrum that is less amenable to filtering.[47] Nevertheless, some broad bandpass filters, sometimes marketed under the label City Light Suppression (CLS) filters,[48] have been designed to sit between the wavelengths of the main light pollution sources and also sit in the blue-green gap of modern LED lamps.

Nebula filters transmit principally the light close to the emission lines arising from diffuse astronomical sources such as H II regions and planetary nebulae. The strongest visible emission lines from such objects are the neutral hydrogen lines at 486 nm (Hβ) and 656 nm (Hα), and the doubly ionised oxygen "O III" lines at 496 and 501 nm (conveniently close to Hβ). A nebula filter seeks to suppress most of the light at other wavelengths, so it maximises the contrast of gaseous sources. Some filters in this category are known as UHC (ultra-high contrast) filters.

Figure 6.7a shows the sensitivity of the human eye under low light levels (scotopic conditions) when it is dark-adapted, and the wavelengths of the major astrophysical emission lines noted above. Figure 6.7b shows competing sources of light such as the sharp emission lines from low-pressure sodium lamps and mercury lamps (mercury lines are also present in fluorescence tube lamps), and the broader emission arising from high-pressure sodium lamps[49] and white LED streetlights.[50] Fortuitously, the Hβ and O III emission lines and the peak of the human scotopic vision band coincide with the gap between the narrow blue and broad green-red emission peaks of modern LED lighting as well as in the gap between emissions from the major older-technology fluorescent tube and sodium street lamps, so existing CLS, UHC and nebula filters should continue to provide some benefit even where LED streetlighting replaces sodium lamps. A wide range of possible filters exists; a small sample of three blue-transmitting filters including an inexpensive dye-based Wratten filter and two more specialist (neodymium and dielectric CLS) filters, is shown in Figure 6.7c.

Whereas observers might find it distracting to observe a bright planet like Jupiter through a coloured filter because of the obvious colour change upon observation, observers of fainter diffuse objects will not be so affronted because faint objects fail to excite the cone cells, and rod cells provide only intensity information to the brain, not colour information. Consequently, a faint nebula or galaxy observed through a blue CLS filter will not appear blue, whereas brighter stars and planets observed through the same filter will; it is then very obvious to the observer when their eyes have transitioned between photopic, cone-facilitated vision and scotopic, rod-facilitated vision.

It may be stating the obvious, but newcomers should note that filters cannot make your target brighter, they can at best improve the contrast if you have unwanted light in the field of view, and ultimately the eye needs contrast to be able to detect features that distinguish one part of a target from another, or to distinguish a target from the background in a light-polluted sky. It is worth remembering that the famous Messier catalogue was compiled in the 1770s and 1780s, using a ~4 inch telescope in pre-revolutionary Paris, but the chances of seeing the Messier objects from central Paris, or indeed most astronomers' backyards, are now much diminished by the proliferation of urban electric street lighting which has robbed them of contrast. Nebula filters and other sky pollution filters try to redress the loss of contrast, but can only do so at the cost of losing further light.

(a)

(b)

(c)

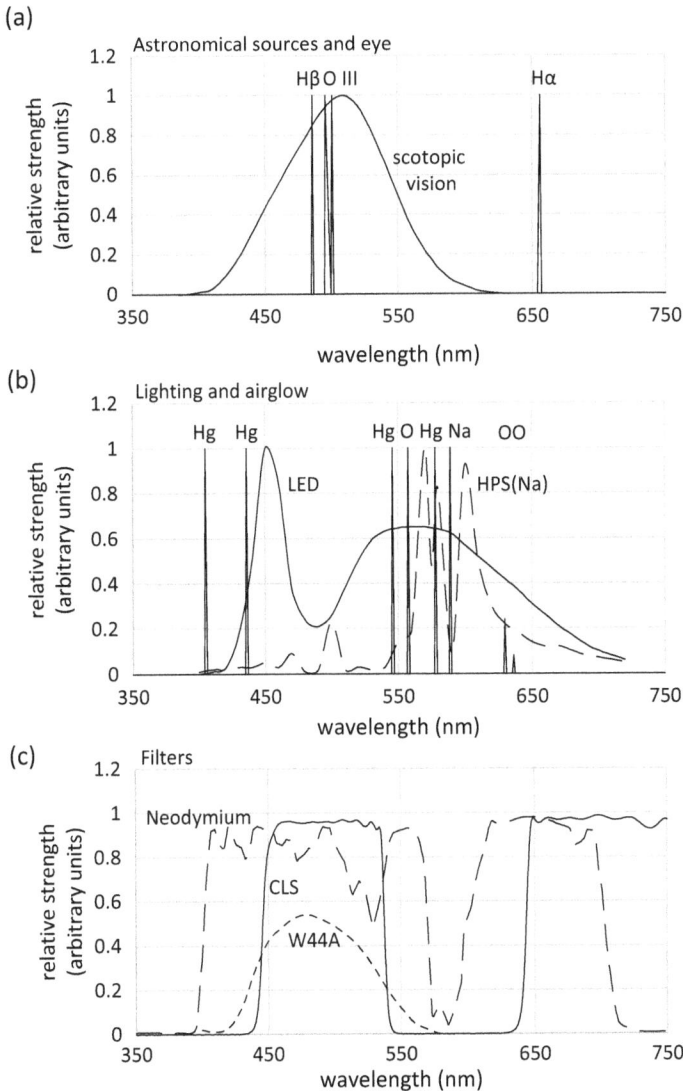

FIGURE 6.7 Spectral curves. (a) Sensitivity of human eye under scotopic (low light) conditions (continuous curve), and relevant H I and O III emission lines seen in astronomical emission line sources (spikes). (b) Unwanted light: white LED street lamp (continuous curve),[51] high-pressure sodium lamps (dashed curve "HPS(Na)"),[52] and key emission lines from low-pressure sodium (Na), mercury-containing lamps (Hg) and airglow from neutral atomic oxygen (O)[53] (spikes). (c) Sample of blue-transmitting filters to restrict the mix of wanted and unwanted wavelengths reaching the observer's eye: neodymium (dashed curve),[54] Wratten 44A (dotted curve)[55,56] and CLS filter (solid curve).[57]

Diagonals

Many portable telescopes are equipped with a mirror or prism that folds the light path through an angle of 90° or similar shortly before it reaches the eyepiece. Their purpose is to make it more convenient to look into the eyepiece compared to a straight-through configuration where an astronomical telescope points up and the eyepiece points down. The two approaches are to use a fully reflecting surface such as a flat mirror, which could have an aluminium coating or more exotic dielectric finish, or a prism whereby total internal reflection occurs off the long (hypotenuse) face.

A clean aluminium surface typically has a reflectance of 85% in the visible spectrum, so this simplest approach comes with some loss of light. More highly reflective finishes such as protected silver or enhanced dielectric coatings can be applied, which can provide a reflectance approaching 99%, at a greater price. Depending on the environmental conditions and the quality of the deposition, specialist dielectric coatings can have a shorter life than aluminium, though modern coating technology has reduced this concern.

If a prism is used, the light enters the prism through a flat face, undergoes total internal reflection at an angled face (45° for a 90° fold of the beam), and then exits through a second flat face. Total internal reflection is 100% efficient (in the absence of contaminants on the hypotenuse face which may vary the refractive index on the back face of the glass), so the light losses associated with a prism diagonal are minimal so long as antireflection coatings are applied to the two flat air-glass surfaces where the light enters and exits (otherwise two 4% losses are incurred there).

It should be noted that, optically, a prism does more than reflect the rays through 90°. The light entering the prism is refracted as it enters the first flat face, travels a distance comparable to the barrel diameter (1¼ or 2 inches) through the glass (n~1.5), and is then refracted again at the face where it leaves the prism. Even though the faces are flat (plano), they are traversed by converging pencils of rays and therefore introduce aberrations that scale linearly with the thickness of the glass.[58] The same is therefore true for light passing through a filter, but a filter is usually so thin (~1 mm thick) that the aberrations are completely negligible in practice. For a 1¼ inch prism operating in a slow optical path, say f/10 or so, the aberrations are also usually sufficiently small compared to those already present in the eyepieces that they too can be ignored, but for a larger (2 inch) prism operating in a faster (say f/6) system, the aberrations can be significant, especially at higher field angles, and longitudinal chromatic aberration in particular may deteriorate. An optical designer could assume that such a prism will always be used with the eyepiece, and optimise the eyepiece accordingly, but it would be difficult to ensure this assumption is always met in use on a telescope where accessories are interchangeable.

The surfaces of mirror diagonals are more likely to produce higher levels of light scatter than prism diagonals, degrading the image by obscuring subtle contrast variations. In such cases where minimal scattering is preferable, it is possible to remove the diagonal entirely and thus eliminate the scattering, but whether doing this is productive or not also depends on the telescope layout. Most Cassegrain-style telescopes are focussed by varying the separation of the primary and secondary mirrors, and designers will generally have assumed the presence of a diagonal assembly in determining where the eyepiece needs to be placed to achieve focus with the primary and secondary mirror at their optimal separation. Removing the diagonal without also inserting a straight-through spacer with the same tube length would alter the eyepiece location considerably and may therefore require the primary-secondary separation to be set to a non-optimal value to place the telescope focal plane at the field stop of the newly positioned eyepiece. This in turn may degrade the imaging performance of the telescope and counteract any improvement achieved by removing the light scatter from the diagonal. An empirical assessment may be required to assess the pros and cons on a particular telescope, with a given diagonal and eyepiece selection.

Binocular Viewers

A binocular viewer is an optical splitter that divides the beam from the telescope into two light paths, each carrying half of the light. The light paths feed two identical eyepieces, which are spaced apart in a similar way to the eyepieces on handheld binoculars and thus allow the observer to use both eyes simultaneously when observing. A similar arrangement has been used on non-stereo binocular microscopes for many decades. Obviously, halving the light is not recommended for very faint objects where the eye is struggling to detect the light at all, but where there is a good level of light available, such as for the Moon, planets, star clusters and bright nebulae, the loss of intensity in each eye is made up for by the ability of the brain to combine the signals from the two eyes into an interpreted image. Provided the binocular viewer is correctly adjusted for the spacing (interpupillary distance) and different powers of the observer's eyes, the brain can better process features common to the optical images incident on both eyes, improving contrast sensitivity and visual acuity.[59] Deficiencies in the image-forming capability of an individual eye can be ignored or corrected by the brain in interpreting the pair of images. Such deficiencies can include short-lived fluctuations in the accommodation, and fine-scale disruptions of the image due to "entopic shadows" (internally-created by the eye) such as due to scattering by small specks floating in the tears that moisten the cornea, and from shadows and diffraction patterns cast onto the retina by floaters (*muscae volitantes*), which are proteins and other cell fragments floating in the vitreous humour (interior fluid) of the eye.[60,61] Floaters may be more obvious during observations of bright planetary targets at high magnification because a small exit pupil is implied, and the smaller width of each pencil of rays leads to more clearly defined floater shadows and diffraction fringes. Under binocular observations, however, the brain can eliminate these defects from the reconstructed mental image that we perceive and appreciate.

Binocular viewers obviously add weight and cost to a telescope setup, once the binocular viewer and a pair of eyepieces are included. Binocular viewers tend to accommodate only 1¼ inch eyepieces, lacking the field of view to justify or accommodate 2 inch eyepieces.

Because of the necessity of setting the interocular distance appropriately and adjusting the relative focus power of the two eyepieces to match the observer's eyes, it is not trivial to adjust a binocular viewer for a number of observers in succession; they are better suited to the activities of a single observer. Observers often cite a more natural, relaxed experience for their eyes when using binocular viewers, compared with the alternatives of either covering a single eye or learning to observe with both eyes open when using a single eyepiece.

This list of eyepiece accessories is far from exhaustive but aims to capture the more common ones that may enhance the visual observing experience if employed in the right way, at the right time, on the right telescope. Not every telescope needs to be equipped with all possible eyepieces and all possible accessories.

I hope that by reading this book, you have gained a better understanding of how eyepieces work, what they are capable of showing you, how to interpret marketing claims in the context of your own specific equipment, and ultimately that you derive more enjoyment from using your eyepieces well to observe the Universe.

NOTES

1 E. Hecht, *Optics*, 5th edition, Pearsons, 2017, Chapter 4.

2 V. Blahnik and B. Voelker, About the reduction of reflections for camera lenses, Carl Zeiss AG, 2016 https://lenspire.zeiss.com/photo/app/uploads/2022/02/technical-article-about-the-reduction-of-reflections-for-camera-lenses.pdf (accessed 09/03/2025).

3 Carl Zeiss, Verfahren zur Erhöhung der Lichtdurchlässigkeit optischer Teile durch Erniedrigung des Brechungsexponenten an den Grenzflächen dieser Teile, German Patent 685767, 1939 https://worldwide.espacenet.com/patent/search/family/007625750/publication/DE685767C?q=DE685767 (accessed 24/04/2025).

4 V. Blahnik and B. Voelker, About the reduction of reflections for camera lenses, Carl Zeiss AG, 2016 https://lenspire.zeiss.com/photo/app/uploads/2022/02/technical-article-about-the-reduction-of-reflections-for-camera-lenses.pdf (accessed 09/03/2025).

5 Anti-reflection (AR) coatings, Edmund Optics® https://www.edmundoptics.com/knowledge-center/application-notes/lasers/anti-reflection-coatings/ (accessed 09/03/2025).

6 Most major amateur astronomy magazines (print and/or online) publish reviews of eyepieces by experienced observers, e.g. https://www.skyatnight-magazine.com/reviews/eyepieces (accessed 09/03/2025).

7 R. Kingslake and R.B. Johnson, *Lens Design Fundamentals*, 2nd edition, Academic Press, 2010 Chapter 3.

8 TeleVue Nagler Planetary Zoom https://www.televue.com/engine/TV3b_page.asp?id=49 (accessed 16/03/2025).

9 SVBONY Planetary Zoom Eyepiece https://www.svbony.com/sv215-planetary-zooms-eyepiece/ (accessed 16/03/2025).

10 APM Super Zoom eyepiece https://www.apm-telescopes.net/en/apm-super-zoom-eyepiece-77mm-to-154mm-with-125-connector-and-filter-thread (accessed 17/03/2025).

11 Neewer Zoom Eyepiece https://uk.neewer.com/products/neewer-ls-t22-zoom-telescope-eyepiece (accessed 17/03/2025).

12 Lunt Zoom Eyepiece https://www.bresseruk.com/p/lunt-ls7–21ze-zoom-eyepiece-7.2mm-21.5mm-0554501 (accessed 17/03/2025).

13 SVBONY Super Zoom Aspheric https://www.svbony.com/sv230-zoom-aspheric-eyepiece/#W9185A (accessed 17/03/2025).

14 Baader planetarium Hyperion Zoom eyepiece https://www.baader-planetarium.co.uk/shop/baader-hyperion-universal-zoom-mark-iv-8-24mm-eyepiece-1-2/ (accessed 17/03/2025).

15 Bresser LER zoom eyepiece https://www.bresseruk.com/p/bresser-ler-zoom-eyepiece-deluxe-8-24mm-1.25–4920320 (accessed 17/03/2025).

16 Celestron® Zoom Eyepiece https://www.celestron.com/products/8-24mm-zoom-eyepiece-125in (accessed 17/03/2025).

17 R. Kingslake and R.B. Johnson, *Lens Design Fundamentals*, 2nd edition, Academic Press, 2010, Chapter 6.

18 R. Kingslake and R.B. Johnson, Lens Design Fundamentals, 2nd edition, Academic Press, 2010, Chapter 15.

19 Multiply Equation 3.2 $\zeta = r - \sqrt{r^2 - y^2}$ by $\dfrac{r + \sqrt{r^2 - y^2}}{r + \sqrt{r^2 - y^2}}$, i.e. by a particularly useful form of "1", multiply out the numerator to find it equals y^2, and take a factor r out the front of the denominator.

20 M. Kidger, *Fundamental Optical Design*, SPIE, 2000, Chapter 1.

21 R. Kingslake and R.B. Johnson, Lens Design Fundamentals, 2nd edition, Academic Press, 2010, Chapter 15.

22 D.J. Schroeder, *Astronomical Optics*, 2nd edition, Academic Press, 2000, Chapter 4.

23 M. Kidger, *Fundamental Optical Design*, SPIE, 2000, Chapter 1.

24 M. Kidger, *Fundamental Optical Design*, SPIE, 2000, Chapter 3.

25 M. Kidger, *Fundamental Optical Design*, SPIE, 2000, Chapter 1.

26 25 mm Dia., 25 mm FL, 532nm V-Coat, Hyperbolic Aspheric Lens, Stock # 89–431 https://www.edmundoptics.co.uk/p/25mm-dia-x-25mm-fl-532nm-v-coat-best-form-aspheric-lens/31938/ (accessed 13/03/2025).

27 S. Sugawara, Eyepiece lens of wide visual field, United States Patent Office, US5684635, 1987 https://worldwide.espacenet.com/patent/search/family/027321729/publication/US5684635A?q=pn%3DUS5684635A (accessed 12/03/2025).

28 M. Kanai, Eyepiece system, United States Patent Office, US5790313, 1998 https://worldwide.espacenet.com/patent/search/family/011520498/publication/US5790313A?q=pn%3DUS5790313 (accessed 14/03/2025).

29 J. Hall, Eyepiece assembly using plastic aspheric element, United States Patent Office, US6282030, 2001 https://worldwide.espacenet.com/patent/search/family/024220886/publication/US6282030B1?q=pn%3DUS6282030 (accessed 14/03/2025).

30 M. Kanai, Eyepiece system, United States Patent Office, US5790313, 1998 https://worldwide.espacenet.com/patent/search/family/011520498/publication/US5790313A?q=pn%3DUS5790313 (accessed 14/03/2025).

31 Hyperion aspheric eyepieces technical data https://www.baader-planetarium.com/en/downloads/dl/file/id/123/hyperion-aspheric-eyepieces-technical-data.pdf (accessed 14/03/2025).

32 Svbony aspheric eyepiece https://www.svbony.com/optional-lens-4mm-10mm-23mm-eyepiece (accessed 14/03/2025).

33 SVBONY Super Zoom Aspheric https://www.svbony.com/sv230-zoom-aspheric-eyepiece/#W9185A (accessed 17/03/2025).

34 Thorlabs Gradient Index (GRIN) lenses for imaging https://www.thorlabs.com/newgrouppage9.cfm?objectgroup_id=11167 (accessed 17/03/2025).

35 J.D. Forer, S.N. Houde-Walter, J.J. Miceli et al., Gradient-index eyepiece design, *Applied Optics*, 22(3), 407–412, 1983 https://doi.org/10.1364/AO.22.000407 (accessed 18/03/2025).

36 Thorlabs Gradient Index (GRIN) lenses for imaging https://www.thorlabs.com/newgrouppage9.cfm?objectgroup_id=11167 (accessed 17/03/2025).

37 Edmund Optics® Gradient index (GRIN) Rod Lenses https://www.edmundoptics.co.uk/f/gradient-index-grin-rod-lenses/13814/ (accessed 17/03/2025).

38 J.D. Forer, S.N. Houde-Walter, J.J. Miceli et al., Gradient-index eyepiece design, *Applied Optics*, 22(3), 407–412, 1983 https://doi.org/10.1364/AO.22.000407 (accessed 18/03/2025).

39 H. Tsuchida, Eyepiece lens, United States Patent Office, US5986817, 1999 https://worldwide.espacenet.com/patent/search/family/015460253/publication/US5986817A?q=pn%3DUS5986817A (accessed 17/03/2025).

40 M.H. Freeman and C.C. Hull, *Optics*, 11th edition, Butterworth Heinemann, 2003, Chapter 15.

41 D.A. Atchison and G. Smith, *Optics of the Human Eye*, 2nd edition, CRC Press, 2023, Chapter 17.

42 W. Crookes, Note on the absorption spectrum of didymium, Nature, 34, 266, 1886 https://www.nature.com/articles/034266b0 (accessed 19/03/2025).

43 W. Crookes, XXVII. The Bakerian Lecture.-On radiant matter spectroscopy: The detection and wide distribution of yttrium, *Philosophical Transactions of the Royal Society of London* 174, 891–918, 1883 https://royalsocietypublishing.org/doi/abs/10.1098/rstl.1883.0027 (accessed 19/03/2025).

44 Didymium safety glasses https://phillips-safety.com/didymium-glasses/ (accessed 19/03/2025).

45 Baader neodymium moon and skyglow filter https://www.baader-planetarium.com/en/baader-neodymium-moon-and-skyglow-filter.html (accessed 19/03/2025).

46 Baader neodymium filters https://www.baader-planetarium.com/en/baader-neodymium-moon-and-skyglow-filter.html (accessed 20/03/2025).

47 C. Elvidge, D. Keith, B. Tuttle and K. Baugh, Spectral identification of lighting type and character, *Sensors*, 10, 3961–3988, 2010 https://doi.org/10.3390/s100403961 (accessed 20/3/2025).

48 Astronomik filters for visual observation https://www.astronomik.com/en/visual-filters.html (accessed 20/03/2025).

49 A. Sánchez de Miguel et al., ISS nocturnal images as a scientific tool against Light Pollution: Flux calibration and colors, Highlights of Spanish Astrophysics VII, 2012, in Valencia, Spain. J.C. Guirado et al. eds. 2012 https://www.researchgate.net/publication/258843567 (accessed 27/03/2025).

50 C. Elvidge, D. Keith, B. Tuttle and K. Baugh, Spectral identification of lighting type and character, *Sensors*, 10, 3961–3988, 2010 https://doi.org/10.3390/s100403961 (accessed 20/3/2025).

51 Ch. Leinert et al., The 1997 reference of diffuse night sky brightness, *Astronomy and Astrophysics Supplement*, 127, 1–99, 1997 https://aas.aanda.org/articles/aas/full/1998/01/ds1449/ds1449.html (accessed 27/03/2025).

52 A. Sánchez de Miguel et al., ISS nocturnal images as a scientific tool against Light Pollution: Flux calibration and colors, Highlights of Spanish Astrophysics VII, 2012, in Valencia, Spain. J.C. Guirado et al. eds. 2012 https://www.researchgate.net/publication/258843567 (accessed 27/03/2025).

53 Ch. Leinert et al., The 1997 reference of diffuse night sky brightness, *Astronomy and Astrophysics Supplement*, 127, 1–99, 1997 https://aas.aanda.org/articles/aas/full/1998/01/ds1449/ds1449.html (accessed 27/03/2025).

54 Baader neodymium filters https://www.baader-planetarium.com/en/baader-neodymium-moon-and-skyglow-filter.html (accessed 20/03/2025).

55 Kodak® Wratten 44A filter https://www.edmundoptics.co.uk/p/cyan-44a-kodak-wratten-color-filter/11205/ (accessed 27/03/2025).

56 Kodak® Wratten-2 filters https://www.kodak.com/en/motion/page/wratten-2-filters/ (accessed 27/03/2025).

57 Astronomik filters for visual observation https://www.astronomik.com/en/visual-filters.html (accessed 20/03/2025).

58 D.J. Schroeder, *Astronomical Optics*, 2nd edition, Academic Press, 2000, Chapter 7.

59 D.A. Atchison and G. Smith, *Optics of the Human Eye*, 2nd edition, CRC Press, 2023, Chapter 1.

60 E. Hecht, *Optics*, 5th edition, Pearsons, 2017, Chapter 5.

61 R.B. Rabbetts, *Bennett and Rabbetts' Clinical Visual Optics*, 4th edition, Butterworth-Heinemann, 2007, Chapter 22.

Index

Note: *Italic* page numbers refer to figures and page numbers followed by "n" refer to end notes.

For Product Safety Concerns and Information please contact our EU
representative GPSR@taylorandfrancis.com
Taylor & Francis Verlag GmbH, Kaufingerstraße 24, 80331 München, Germany

www.ingramcontent.com/pod-product-compliance
Lightning Source LLC
Chambersburg PA
CBHW060554220326
41598CB00024B/3104